PRAISE FOR *THE FLY IN THE OINTMEN*

"Joe Schwarcz has done it again. In fact every bit as entertaining, informative, a celebrated collections, but contains enric [illegible] percent more attitude per chapter. Whether he's assessing the legacy of Rachel Carson, coping with penile underachievement in alligators, or revealing the curdling secrets of cheese, Schwarcz never fails to fascinate."

—Curt Supplee, former science editor, *Washington Post*

"Wanna know how to wow 'em at a cocktail party or in a chemistry classroom? Take a stroll through the peripatetic journalistic world of the ideas and things of science brought to life by Dr. Joe. Here is narrative science at its best. The end result? In either place, scientific literacy made useful by *The Fly in the Ointment*."

—Leonard Fine, professor of chemistry, Columbia University

PRAISE FOR *DR. JOE AND WHAT YOU DIDN'T KNOW*

"Any science writer can come up with the answers. But only Dr. Joe can turn the world's most fascinating questions into a compelling journey through the great scientific mysteries of everyday life. *Dr. Joe and What You Didn't Know* proves yet again that all great science springs from the curiosity of asking the simple question . . . and that Dr. Joe is one of the great science storytellers with both all the questions and answers."

—Paul Lewis, president and general manager, Discovery Channel

PRAISE FOR *THAT'S THE WAY THE COOKIE CRUMBLES*

"Schwarcz explains science in such a calm, compelling manner, you can't help but heed his words. How else to explain why I'm now stir-frying cabbage for dinner and seeing its cruciferous cousins—broccoli, cauliflower, and brussels sprouts—in a delicious new light?"

—Cynthia David, *Toronto Star*

PRAISE FOR *RADAR, HULA HOOPS, AND PLAYFUL PIGS*

"It is hard to believe that anyone could be drawn to such a dull and smelly subject as chemistry—until, that is, one picks up Joe Schwarcz's book and is reminded that with every breath and feeling one is experiencing chemistry. Falling in love, we all know, is a matter of the right chemistry. Schwarcz gets his chemistry right, and hooks his readers."

—John C. Polanyi, Nobel Laureate

Is That a Fact?

Also by Dr. Joe Schwarcz

The Right Chemistry: 108 Enlightening, Nutritious, Health-Conscious, and Occasionally Bizarre Inquiries into the Science of Everyday Life

Dr. Joe's Health Lab: 164 Amazing Insights into the Science of Medicine, Nutrition, and Well-Being

Dr. Joe's Brain Sparks: 179 Inspiring and Enlightening Inquiries into the Science of Everyday Life

Science, Sense & Nonsense: 61 Nourishing, Healthy, Bunk-Free Commentaries on the Chemistry that Affects Us All

Brain Fuel: 199 Mind-Expanding Inquiries into the Science of Everyday Life

An Apple a Day: The Myths, Misconceptions and Truths About the Foods We Eat

Let Them Eat Flax: 70 All-New Commentaries on the Science of Everyday Food & Life

The Fly in the Ointment: 70 Fascinating Commentaries on the Science of Everyday Life

Dr. Joe and What You Didn't Know: 177 Fascinating Questions and Answers about the Chemistry of Everyday Life

That's the Way the Cookie Crumbles: 62 All-New Commentaries on the Fascinating Chemistry of Everyday Life

The Genie in the Bottle: 64 All-New Commentaries on the Fascinating Chemistry of Everyday Life

Radar, Hula Hoops, and Playful Pigs: 67 Digestible Commentaries on the Fascinating Chemistry of Everyday Life

IS THAT A FACT?

Frauds, Quacks, and the Real Science of Everyday Life

Dr. Joe Schwarcz

ECW PRESS

Published by ECW Press
2120 Queen Street East, Suite 200
Toronto, Ontario, Canada M4E 1E2
416-694-3348 / info@ecwpress.com

LIBRARY AND ARCHIVES CANADA CATALOGING IN PUBLICATION

Schwarcz, Joe, author
Is that a fact? : frauds, quacks, and the real science of
everyday life / Dr. Joe Schwarcz.

Includes index.
Issued in print and electronic formats.
ISBN 978-1-77041-190-6 (PBK.). ISBN 978-1-77090-527-6 (PDF).
ISBN 978-1-77090-528-3 (EPUB)

1. Science—Popular works. 2. Science—Miscellanea. I. Title.

Q173.S385 2014 500 C2013-907759-6 C2013-907760-X

Cover design: David Gee
Cover images: rubber duck: Colevine yard / iStock; beaker: malerapaso / iStock
Typesetting and production: Lynn Gammie
Printing: Friesens 1 2 3 4 5

The publication of *Is That a Fact?* has been generously supported by the Canada Council for the Arts which last year invested $157 million to bring the arts to Canadians throughout the country, and by the Ontario Arts Council (OAC), an agency of the Government of Ontario, which last year funded 1,681 individual artists and 1,125 organizations in 216 communities across Ontario for a total of $52.8 million. We also acknowledge the financial support of the Government of Canada through the Canada Book Fund for our publishing activities, and the contribution of the Government of Ontario through the Ontario Book Publishing Tax Credit and the Ontario Media Development Corporation.

PRINTED AND BOUND IN CANADA

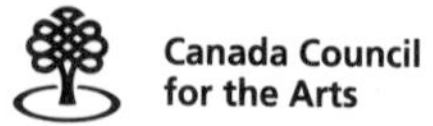

Canadä

TABLE OF CONTENTS

IN THE BEGINNING

BLACK

GRAY

WHITE

IN THE END

IN THE BEGINNING

Is That a Fact?

"Is that a fact?" "They say that . . ." "I heard that . . ." Just listen in on a few conversations around the water cooler and it won't be long before one of these phrases rings out. After all, this is the Communication Age. We are connected through cell phones, radio, TV, and, of course, the web. We talk, we Tweet, we link, we text, we Facebook. We are informed. But in many cases, unfortunately, we are also misinformed.

We suffer from information overload. Just Google a subject and within a second, you can be flooded with a million references. It is therefore more important than ever to be able to analyze those references and know how to separate sense from nonsense. And that's where learning comes in. Information has to be scrutinized in the light of what is already known. But learning must be coupled with critical thinking. Confucius said it very well: "Learning without thought is labor lost; thought without learning is perilous."

The University of Google is well stocked with information, but its students are left to flounder when it comes to determining whether that information is reliable. Accounts of miraculous cancer cures, the rants of anti-vaccine activists, the exploits of so-called psychics, and the claims of various alternative healers may sound very seductive, but stand to lose their luster in the light of scientific education. It would, however, be incorrect to suggest that education is the vaccine against folly. The annals of history are replete with examples of educated people who have succumbed to nonsense. Sir Arthur Conan Doyle, a physician by training, believed in fairies and in communicating with the dead. Curiously, he was the creator of Sherlock Holmes, who was a logician extraordinaire and eschewed such silliness.

Indeed, it was Holmes who reminded us, "It is a capital mistake to theorize before one has data. Insensibly one begins to twist facts to suit theories instead of theories to suit facts." These days, those of us who follow Holmes's dictum and put evidence-based science on a pedestal often get criticized for challenging claims we consider to be unscientific. "They laughed at Galileo," the promoters of such claims say, "and at Columbus, and at the Wright Brothers." But, as Carl Sagan pointed out, the fact that some geniuses were laughed at does not imply that all who are laughed at are geniuses. They also laughed at Bozo the Clown.

Our best bet in order to differentiate the Bozos from the prospective Galileos is to push for more science education at all levels, with a strong emphasis on the importance of critical thinking. Furthermore, it should be realized that when it comes to separating sense from nonsense, mental prowess is not enough. Benjamin Franklin was right on when he opined, "Genius without education is like silver in a mine." Indeed, the value is there, but the silver is not of much use until you extract it. But how do you go about this extraction? How do we know who is right and who is wrong? How do we know what is a fact and what is not? How do I know what I claim to know?

Actually, that is a question I had to contemplate recently when a student innocently asked me, "And how do you know that?"

I had just finished a lecture on toxicology in which I had described the problem of cyanide poisoning by cassava, a tuber similar to the potato that is a staple in some parts of Africa. However, with some varieties of cassava, there's an issue: if not properly processed, it can harbor a lethal amount of cyanide. (This is not the case with the cassava grown in the Caribbean.) But soaking the peeled tuber in water for several days releases enzymes that degrade the cyanide-storage compound linamarin, causing the toxic cyanide to be dissipated into the air as hydrogen cyanide. Unfortunately, cases of acute cyanide poisoning have occurred when famine conditions forced a shortening of the soaking time. Since even proper processing doesn't remove all the cyanide, chronic low-level exposure can lead to goiter or even konzo, a type of paralysis.

I've described the cyanide connection in lectures numerous times, but never before had I been asked a question about how I had acquired this knowledge. It did start me thinking. Indeed, I've never been to Africa, have never even seen a live cassava plant. I've never carried out any testing of cassava for cyanide. Truth be told, I wouldn't even know how to go about it, although I think that with a little digging, I could figure it out. I do have a vague recollection of once eating fried cassava somewhere in the Caribbean, but that's as close as I've come to experimenting with the tuber. So, in fact, how do I know about its chemistry? It all comes down to reading various accounts of cassava poisoning in toxicology and chemistry texts.

And how do the authors of these texts know what they are writing about? Chances are they haven't had any closer encounters with cassava than I have. But they have read the peer-reviewed literature on the topic, have digested the facts, and have managed to piece together the story. They would have read a paper in a medical journal about how the symptoms of konzo were traced to cyanide poisoning and about how a link to cassava was discovered. Then, in a chemical publication, they would have learned that the actual culprit, linamarin, is present in unprocessed cassava but not in the soaked version. Finally, a paper likely published in a biochemistry journal would have revealed the action of enzymes on linamarin. Basically, then, what we call scientific knowledge is gained through a distillation of the relevant peer-reviewed literature. And that literature is the altar at which scientists worship. But, as with religion, there is faith involved. Faith that the peer-reviewed literature can be trusted. That faith, however, cannot be blind. It must be tempered with a dose of skepticism.

So how does the peer-review process work? A principal investigator (PI), who may be an academic, industrial, or government researcher, designs a study, let's say on how a novel weight-reducing drug affects mice. The work may be carried out by himself or by other members of his research group. He or she then writes a paper with the results, adds an appropriate discussion, and submits it to a journal that is geared toward such subject matter. The journal's editor, who has a general command of the science normally featured in the publication, then sends the paper on to two or three referees who have expertise in the specific research area in question. These referees, usually researchers themselves, critique the paper and often ask for clarification or even for more work to be done. The paper then goes back for comments to the original author, who is unaware of the identity of the referees. This process can go back and forth several times before a paper is either accepted for publication or is rejected. Once published, other scientists may weigh in with their opinions or criticisms, which then might appear in subsequent issues as letters to the editor.

Some researcher may, upon reading the paper, wish to extend the work, perhaps by mounting a human trial of the drug. First, though, repetition with more animals may be in order. If the repetition is successful, the drug starts to get more traction and invites further research. By the time it is approved for human use, it will have been the subject of a good number of peer-reviewed papers. Then we can say "we know" it works, albeit with some apprehension.

Why apprehension? Because the peer-review process is not perfect. First, the referees of course cannot repeat the work, which is often the result of years of research. They have to assume that what the author says was done really was done, that it was done well, and that the results have been accurately reported. The PI has to assume the same as far as his research group goes. But humans are, well, human. Some work may be sloppy, and results that do not seem to "fit the curve" may be deemed to be erroneous and therefore ignored. There may also be discrepancies or outright fraud that are not detected until years after a paper has been published. A case in point is Andrew Wakefield's infamous publication in *The Lancet* suggesting a link between autism and vaccination. Twelve years passed before it became clear that the work could not be reproduced, prompting the journal to withdraw the paper, noting that "elements of the manuscript had been falsified." By that time, an increase in measles fatalities attributed to a decrease in vaccination rates had already been noted.

Problems may eventually crop up even with research that was properly carried out. A side effect of a medication that affects a fraction of a percent of patients will not be detected in trials, but will become obvious when millions take the drug.

So peer-review isn't the end-all. But remember what Churchill said about democracy? "It is the worst form of government except for all the others that have been tried." Ditto for the peer-review process. Peer review, however, is the final stage in a scientific investigation that usually begins with an observation that prompts a comment along the lines of "gee, that's funny." And that observation may happen in a serendipitous fashion. But in the words of Louis Pasteur, "Chance favors the prepared mind." That oft-quoted expression is a great springboard for our dive into the pool of science.

Chance Favors the Prepared Mind

I had my tonsils removed in 1954. In those days, a few bouts of tonsillitis, and out they came. I remember being plied with chloroform before the operation and with ice cream after. I also remember being given a special gum, "imported from America," to chew. It was probably some version of Aspergum, which contained aspirin and was supposed to relieve the sore throat. The idea of using the gum after a tonsillectomy was introduced in the 1940s by Lawrence Craven, a California physician, who made an interesting observation: patients who chewed the gum bled more, leading Craven to speculate that aspirin had an anti-clotting effect. It was already known at the time that heart attacks and strokes could be caused by blood clots, and Craven began to treat his adult coronary disease patients with aspirin. He noted a reduced frequency of heart attacks! Craven published his findings, but because he had no controls, they were mostly ignored until British biochemist John Vane clearly demonstrated aspirin's effect on the blood in 1971. Today, aspirin is standard therapy for people at risk for cardiovascular disease, tracing back to Lawrence Craven's serendipitous finding.

The word "serendipity" was introduced into the English language in the eighteenth century by writer Horace Walpole, who was taken by the ancient Persian tale of the "Three Princes of Serendip," who during their travels made a number of discoveries "by accidents and sagacity of things they were not in quest of." "Serendipity" came into common use as a description of a "lucky turn of events," and Walpole's original link to sagacity, defined as "penetrating intelligence, keen perception, and sound judgment" was ultimately forgotten. Walpole's intent was to convey the idea that an accidental discovery doesn't amount to much if the discoverer is not astute enough to capitalize on the chance finding.

The three princes of Serendip certainly exhibited sagacity after accidentally coming on some strange animal tracks on a road. When they later learned from a merchant that he had lost a camel, the princes give him a remarkable description of the animal. "The camel is lame, blind in one eye, is missing a tooth, carried honey on one side and butter on the other, and was ridden by a pregnant woman." When asked how they could possibly have come up with such an accurate description, the princes explained that grass had been eaten from the side of the road where it was less green, so the camel was blind on the other side. Because there were lumps of chewed grass on the road the size of a camel's tooth, the princes inferred they had fallen through the gap left by a missing tooth. The tracks showed the prints of only three feet, the fourth being dragged, indicating that the animal was lame.

The fact that butter was carried on one side of the camel and honey on the other was evident because ants had been attracted to melted butter on one side of the road and flies to spilled honey on the other. There was also an imprint in the dirt from which they deduced the camel had knelt to let down a rider. And why was the rider a pregnant woman? There was some urine nearby, along with some handprints that suggested a woman had needed to use her hands to get up after urinating, her extra weight requiring a push. Shades of Sherlock Holmes. Maybe Sir Arthur Conan Doyle had serendipitously read about the princes of Serendip.

The Persian tale may be somewhat far-fetched, but the story does make a point. The three princes of Serendip were able to capitalize on their chance observation when they heard about the lost camel. And talking about chance, let's return once more to Louis Pasteur's famous comment that "In the field of observation, chance favors the prepared mind."

Pasteur himself furnished a great example of a serendipitous discovery. By 1878, he had formulated his germ theory of disease and had turned his attention to chicken cholera, a problem that plagued the French poultry industry. He managed to isolate a microbe from sick chickens he believed caused the disease and showed that injecting it into healthy birds led to their demise within a day. Scientific evidence requires repetition of an experiment, but a summer vacation intervened. No problem, Pasteur thought, he would just store his bacterial culture. To his astonishment, injecting the culture that had been stored for three months had no effect on the chickens!

He tried again with a fresh culture, and the chickens remained disease free. While many would have concluded that in the original experiment the chickens must have been affected by something other than the suspect bacteria, Pasteur hypothesized that perhaps storage for three months had altered the microbes in a way that resulted in offering protection against infection by fresh bacteria. As it turned out, Pasteur had managed to immunize the chickens with an attenuated microbe! It didn't take long to prove that a weakened form of an infectious organism could impart immunity against the disease normally caused by a more vibrant version. The French chemist then went on to produce vaccines against anthrax and rabies, laying the foundation for the science of immunology, all because his mind was prepared to exercise sagacity when his chickens serendipitously survived an injection of a supposedly deadly microbe.

One of the most famous drugs in the world is also the result of serendipity. Witty advertising and a clever name conjured up to suggest power (from "vitality" and "Niagara") have helped make Viagra a bestseller. Of course, it helps that the drug actually works. But Viagra did not start out life as a treatment for erectile dysfunction. That was a serendipitous finding. The little blue pill was first developed by the Pfizer pharmaceutical company as a possible treatment for angina. In clinical trials, the effects on the heart were less than heartening, but some male patients began to report a surprising uplifting effect. Pfizer researchers were perceptive enough to recognize that they had stumbled upon a potential gold mine, and managed to introduce Viagra to the marketplace within six years, where it has enjoyed stirring serendipitous success in spite of stiff competition.

The Chemistry of Our World Is Too Complex to Be Simplified

We live in a large chemistry lab. A *very* large one. It's called the universe. It may not have shelves stocked with neatly labeled bottles, but everything in it is made up of chemicals. Including us. Indeed, the human body is nothing but a large bag of chemicals — thousands and thousands of them. And they are constantly engaged in all sorts of reactions, which, taken together, constitute life. Amino acids join to make proteins, glucose is "burned" to produce energy, DNA instructs cells to make enzymes, neurotransmitters are synthesized, hormones are cranked out, toxins are eliminated, and thousands of other processes churn out a stunning array of biochemicals necessary to our survival.

It stands to reason that, when dealing with such a complex system, sometimes a wrench gets thrown into the works. It may be a photon of ultraviolet light that causes a break in a strand of DNA, a virus that takes over a cell's machinery, a compound that disrupts hormonal function, a bit of pollen that triggers inappropriate immune activity, a bacterium that spews out toxins, a metal ion that poisons nerve cells, or a chemical that causes cells to multiply irregularly. In fact, the more one learns about the goings-on in the body, and about everything that *can* go wrong, the more remarkable it becomes that anyone is ever healthy.

While sometimes the wrench thrown into the works can be identified, in most cases the specific trigger for a health calamity remains a mystery. Colds can be traced to a virus, an allergy may be pinpointed, and a bout of food poisoning may be linked to a bacterium. But it's another story to determine the cause of damage to a molecule of DNA that resulted in cancer decades later.

Was it a trace of aflatoxin on a moldy peanut? Benzopyrene on that charred steak? Acrylamide in the potato chips? Arsenic in rice? Formaldehyde in cosmetics? Radon seeping into the basement? Or could it have been estrogen in the birth control pills? Aristolochic acid in a dietary supplement? Asbestos in the insulation? Diesel exhaust on the street? Benzene from gasoline fumes? Nitrosamines in hot dogs? Soot from the fireplace? Chloroform in tap water? Naphthalene in mothballs? Phenylenediamine in hair dye? Pesticide residue on an apple?

Or was the cancer triggered by *Helicobacter pylori* bacteria or Human papillomavirus? One could go on and on because there are numerous substances, both natural and synthetic, that can wreak havoc with our biochemistry.

Usually, it is synthetic substances that get blamed, despite the fact that of the sixty million or so known chemicals in existence, only about one tenth of one percent are synthetic. Yet these are the ones that get most of the attention, and usually in a negative way. We hear about "toxic chemicals" and "poisonous chemicals," usually in reference to pesticides, plastic components, cleaning agents, or cosmetic ingredients. Of course, any chemical can be toxic depending on the extent and type of exposure, be it synthetic or natural. The most potent toxin known is botulin, produced by the *Clostridium botulinum* bacterium. A few billionths of a gram, way too small to be seen, can be lethal. Ditto for ricin found in castor beans. All it takes to put you six feet under is an amount equivalent to the weight of half a grain of sand.

The fact is that we live in a fascinatingly complex chemical world. Smell that cup of coffee and you are sniffing hundreds of compounds! A whiff in the bathroom will add about three hundred, many of them, such as methyl mercaptan and skatole, decidedly unpleasant. A single meal will dump thousands and thousands of chemicals into your body, ranging from the proteins, sugars, and fats that plants produce to allow their growth and development to the pigments and scents they use to attract pollinators. Add to this the vast array of compounds plants use to ward off predators. Indeed, we encounter far more natural pesticides than synthetic ones. We are also exposed to a huge array of chemicals produced by industry such as solvents, dry-cleaning compounds, degreasers, paints, plastic additives, pesticides, and packaging materials.

Just to present a picture of chemical diversity and complexity, consider something as simple as honey. Everyone knows that basically it is composed of sugar and water. But "sugar" is a general term for a variety of simple carbohydrates, the most familiar of which are sucrose, glucose, and fructose. But these are not the only sugars found in honey — not by a long shot. There's a long list of others that includes raffinose, gentiobiose, maltose, maltulose, kojibiose, nigerose, and turanose. Then there are proteins, amino acids, and various enzymes that include invertase, which converts sucrose to glucose and fructose, and amylase, which breaks starch down into smaller units. There's also glucose oxidase, which converts glucose to gluconolactone, which in turn yields gluconic acid and hydrogen peroxide. Catalase breaks down the peroxide formed by glucose oxidase to water and oxygen.

Honey also contains trace amounts of the B vitamins riboflavin, niacin, folic acid, pantothenic acid, and vitamin B6. It also has ascorbic acid (vitamin C), and the minerals calcium, iron, zinc, potassium, phosphorous, magnesium, selenium, chromium, and manganese. Then, depending on what plants the bees have been visiting, there are all sorts of flavonoids, of which one, pinocembrin, is unique to honey and bee propolis. There's still more. Honey contains organic acids such as acetic, butanoic, formic, citric, succinic, lactic, malic, and pyroglutamic acids. Use the honey to make cake, and you'll be generating dozens of more compounds, including hydroxymethylfurfural, a potential carcinogen!

By now you are asking yourself if there is a point to this onslaught of chemical terms. There is. It has to do with yet another chemical. And that is bisphenol A (BPA), the plastic component that is pilloried on a daily basis. It is said to cause reproductive problems, heart disease, breast and prostate cancer, brain tumors, obesity, thyroid problems, metabolic syndrome, sexual dysfunction, miscarriage, disruption of dopamine activity, and impairment of fetal development. Now, here is my question: how can this one substance, which chemically resembles so many of the thousands and thousands of compounds to which we are exposed on a regular basis, be responsible for all these horrors? I suppose it is possible, but I doubt that it is likely.

Nevertheless, California, under its brilliant Proposition 65, is set to declare BPA a reproductive hazard based on effects attributed to high doses in animals. Such doses are higher than what the average consumer is exposed to, meaning that there most likely will not be any warning labels. But why don't natural estrogenic compounds raise the same alarm? After all, there are some forty-three foods in the human diet that can be shown to be estrogenically active in the lab. Soybeans, hops, sesame seeds, chickpeas, safflower oil, ginseng, parsley, garlic, and wheat are among them. Considering the extensive use of soy oil, with its significant content of genistein and daidzen, shouldn't there be warnings on foods that contain this oil? How about all products made with soy protein or soy flour? What about cottonseed oil, which has anti-spermatogenic properties and is used in some processed foods?

Certainly, the natural estrogenic compounds have a weak potency when compared with synthetic estrogen. But the same can be said for synthetic compounds with estrogen-like activity. In fact, DDT has a smaller relative estrogen potency than many natural estrogens. When all the natural and synthetic estrogenic compounds to which we are exposed are totaled, estimates are that the estrogenic effects from plant-based estrogens are some forty million times greater than from synthetic chemicals. Of course, these exposures are not mutually exclusive, so just because we are exposed to natural estrogens doesn't mean that we should pile on the synthetics because, after all, the dose makes the poison. But when we compare numbers, we really aren't piling on; we are adding a very small percentage. And something else that doesn't seem to get much consideration: why is it assumed that hormone-like activity is necessarily harmful? Should not the hormesis effect, the possibility of very different effects at lower dose versus the effects at a higher dose, be considered?

I would suggest that there are numerous compounds, both natural and synthetic, that, if studied with as much vigor as BPA, would raise similar concerns. The problem may not be BPA as much as the zeal with which some researchers attempt to convert an association into causation to fit an ideological agenda while ignoring the fact that none of the more than 6,000 studies of BPA has shown that it causes harm to the average consumer. Of course, none of these studies have compared the health of an experimental group exposed to trace amounts of BPA over several decades to a control group with no exposure to BPA but identical exposure to all other chemicals. It hasn't been done because it cannot be done. The chemical complexity of our world is just too great. But that doesn't stop some people from claiming to know things that cannot be known. As physicist Stephen Hawking once said, "The greatest enemy of knowledge is not ignorance, it is the illusion of knowledge."

Callers Have Questions . . .

Time doesn't just fly; it rockets! Seems like yesterday, but it was more than thirty years ago that I started answering callers' questions on the radio. Lots of questions! My rough guess is that by now I've addressed somewhere around 15,000 of them. And recently I had a very interesting one. "What," the caller queried, "was the most bizarre question you've ever been asked?" Well, the truth is that there are lots of candidates for that title, but I think I can narrow it down to a few.

A professor of radiology wondered why noiseless flatus ("silent but deadly") is so much more malodorous than its noisier counterpart. And the answer is . . . I don't know. I wouldn't even hazard a guess without knowing that this is actually true. There do not seem to be any properly controlled studies that have examined this issue, in spite of its obvious importance, and it's not a challenge I'll take up.

People do seem to be fascinated by bodily emissions, both of the human and animal varieties. A caller wanted to know if Canada Geese poop when they fly. He was a golfer who often saw flocks pass overhead and was concerned about looking skyward to follow his ball's flight. The toilet habits of these birds actually have been studied, and it seems they let loose every six to eight minutes, up to some ninety times a day. Although photos of these flying poop machines dropping their bombs are rare, there are plenty of accounts of people suddenly feeling a plop.

Spiders also intrigue people. One young caller wanted to know why the spider he had put in a microwave oven survived. All I can say is that was one lucky spider. Microwave ovens have "cold spots," which is the reason they are usually equipped with turntables. Had the spider climbed aboard that merry-go-round, well, let's just say the ending would not have been merry. When I asked the young investigator how he rewarded his subject for taking part in this experiment, he informed me that he let the arachnid loose in his sister's room. Spiders, he told me, cause high-frequency sound waves to be generated by sisters.

And then there are the seriously strange questions. A graduate student was having a little dispute with his girlfriend, who had come home from massage school with some new information acquired from a classmate. The classmate had recommended she stop eating pork, because pigs have no sweat glands and thus don't sweat out toxins! The specific toxins cruising inside pigs were not mentioned. In any case, the question was whether there was any truth to all of this. Actually, there is some. Pigs really don't sweat. As far as the rest goes, however, it's pure bunk. Sweating does not remove toxins; its purpose is to cool the body through evaporation. Detoxification is a job for the liver and kidneys, organs that do a great job. Sweat glands need not apply.

But if pigs can't sweat, why do we have the expression "sweat like a pig"? That expression is actually derived from the iron-smelting process in which hot iron poured on sand cools and solidifies. The resulting pieces are said to resemble a sow and piglets. Hence, "pig iron." As the pigs cool, the surrounding air reaches its dew point, and beads of moisture form on the surface of the pigs. "Sweating like a pig" indicates that the pig has cooled enough to be safely handled. This pig sweats, but you wouldn't want to eat it.

Now, for the winner! Let me first set it up with a note about phthalates. They're compounds used as plasticizing agents to make certain plastics soft and pliable. Vinyl records, for example, are hard, whereas vinyl shower curtains are soft. The difference is the inclusion of phthalates. While there is no doubt that phthalates are very useful, some are also mired in controversy, mostly because of some widely publicized research showing that pregnant rodents treated with phthalates give birth to male offspring with a shorter-than-normal distance between their anus and their genitals. Such "endocrine-disrupting effects" have led to the restricted use of certain phthalates in items such as vinyl toys that children may put in their mouths.

And the winning question is . . . "Could there be a problem from phthalates leaching out of vinyl electrical tape worn on the skin for years, but changed several times a day?" Now, I've heard of electrical tape being used to wrap hockey sticks and bicycle handlebars, but body parts? That was a new one for me. Especially when I learned what particular body part was being wrapped. It seems the gentleman who had posed the question suffered from "a little urinary incontinence" and discovered that vinyl electrical tape was just the right item to stop any leakage. Having read about the phthalate issue, he became concerned about these chemicals ending up in his blood. The prospect of his ano-genital distance being shortened was not a happy one.

Whether using electrical tape in this inventive fashion constitutes a risk is impossible to say. There are many varieties of tape manufactured with different plasticizers, not always phthalates. Also, there are many phthalates, some of which can pass through the skin, some of which cannot. On top of it all, there is no conclusive evidence that phthalates cause harm to humans. But the ultimate reason this question can't be answered is that nobody has ever carried out a controlled study to examine the effects of wrapping this particular type of wire (perhaps "cable" may be the preferred expression) in electrical tape over lengthy periods, nor do I suspect anyone ever will.

Being on the radio for over thirty years has certainly been fun and educational. From week to week, you just never know what interesting questions will arise. Of course, I can't guarantee answers. Sometimes there just aren't any. Such as when a caller wanted to know what the temperature would be tomorrow if it was zero degrees today and tomorrow promised to be twice as cold. Hmmm. But most of the time, whatever the subject matter, questions tend to boil down to "Is that a fact?" Unfortunately, in science absolute facts are hard to come by.

We race toward them, but the finish line always seems to be receding. Certainty is elusive. Facts are supposed to be based on evidence, but the problem is that with more research, evidence sometimes changes. The fact is that science isn't necessarily white or black; it can come in various shades of gray. And that's just how I've tried to categorize the topics you are about to dive into. "White" entries are factual, at least as far as current evidence indicates, "grays" are a blend of facts and falsehoods, and "blacks" are pretty dismal when it comes to facts. It should all amount to a colorful experience! Let's start by casting some light into the dark shadows of quackery.

BLACK

Quackery Is Not an Issue to Duck

I collect ducks. They're mostly yellow, but the reason I collect them is that they have a figurative "black" side. Ducks quack. And quackery fascinates, amuses, and above all, disturbs me. Of course, the kind of quackery I'm talking about doesn't refer to ducks, but to a special breed of ignorant pretenders to knowledge, usually, but not necessarily, in matters related to health. Quacks boast of providing cures that sound wondrous, but turn out to be no more than clever deceptions.

The origin of the term "quack" is somewhat murky, but may well have been inspired by the resemblance between the sound a duck makes and the rapid-fire oratory of charlatans plying their nostrums. Alternatively, the term may derive from the archaic Dutch expression "*quacksalber*," referring to itinerant mountebanks who hawked various salves that were promoted as having miraculous properties but fell well short of the claims.

The eighteenth and nineteenth centuries are often regarded as the golden age of quackery, but today's peddlers of sham don't have to take a backseat to the colorful schemers of previous eras. Aided and abetted by the internet, they effectively turn gullibility and illness into healthy profits with a mishmash of pseudoscience and seductive testimonials.

The hallmarks of quackery include the bashing of conventional medicine, the use of gushing testimonials from supposed contented patients, and extravagant claims for painless cures for virtually all diseases. For example, "Dr." Brodum's Nervous Cordial and Botanical Syrup, introduced in 1801, was good for "excruciating rheumatic pains and contraction of the joints," as well as for curing the "indiscretions of youth." Questionable qualifications are also often part of the package. Brodum's bogus medical degree was from the Marischal University of Aberdeen, easily confused with the legitimate University of Aberdeen.

Potters's Vegetable Catholicon cured diseases of the liver and "debility resulting from intemperance and dissipation." General Augustus J. Pleasonton promoted the idea that light rays filtered through a special blue glass could arrest disease and restore health. Then there was "Dr." Edrehi's Amulet, containing a berry that released a scent "preventative of fevers and general decline of the system." As a bonus, the berry's strong odor would protect clothing from moths. Quacks also sold intriguing devices, such as Dr. Hawley's mechanical treatment for impotence, cleverly named "The Erector."

While most quack remedies were devoid of biologically active ingredients, James Morison's Vegetable Universal Medicines contained aloes, jalap, gamboge, colocynth, and rhubarb, all of which are plants with laxative properties. Morison invented the "Hygeian System," based on the notion that all pain and disease arise from impurities in the blood and the only effective way of eliminating these impurities is with vegetable purgatives. Calling himself "The Hygeist," he tirelessly attacked the medical profession as the enemy, claiming, "the old medical science is completely wrong," all the while ignoring the potentially lethal effects of the large number of pills he recommended, which of course were only available from his agents.

Scientific advances often foster the promotion of quack cures. William Radam used the discoveries of Robert Koch and Louis Pasteur to promote his "Microbe Killer," which was nothing other than a useless dilute solution of hydrochloric and sulfuric acids. The introduction of electricity launched a variety of "electro galvanic belts" that were to be worn around the middle to cure "nervous debility, female complaints, catarrh, and diseases of the blood." The belts contained copper and zinc disks that constituted a battery capable of producing a mild burning sensation that indicated to the wearer that something was happening. Something was indeed happening, but it had nothing to do with curing any disease. And then there was the "Health Jolting Chair," which targeted the fairer sex, "members of which are particularly prone to neglect the taking of proper exercise and consequently are robbed of the sweet pure breath, vigorous mental action, and vivacious manner characteristic of healthy young womanhood." The chair was equipped with machinery that would provide the jolts that "exercised the internal organs of the body necessary to health."

So that was then. How about now? Well, you can buy a "Quantum Balance Crystal" that will restore the light frequencies missing within our quantum energy field due to the "noxious energy beam that emanates from digital TVs." You can also invest in the "Photon-Genie," which can "devitalize pathogens and detoxify the body with nourishing photobionic energy effectively delivered by both an ionized noble gas energy transmission and deeply penetrating mega frequency life-force energy waves." The babble may have been updated, but nonsense is nonsense, even if it is cloaked in the garb of science.

And did you know that pine pollen can "elevate sexual

libido" (is there any other kind?) and can also "increase fertility and decrease the symptoms of aging"? Or that wearing tourmaline infrared ray socks can improve blood circulation, increase metabolism, and enhance general health? How about fucoidan? This long-chain carbohydrate isolated from seaweed makes "cancer cells self-destruct in as little as seventy-two hours, so cancer cells die by the thousands while healthy cells remain untouched." Of course, this is all hushed up by an evil pharmaceutical industry, as is the "fact" that an extract of the rain forest fruit graviola is 10,000 times as strong as a common chemotherapy drug.

For more mind-numbing claptrap, how about "all-natural liquid oxygen" drops under the tongue to fight jet lag, fatigue, hangover, and aging skin? And while you're at it, you might want to remove the worm-infested toxic sludge from your bowels, a consequence of a "diet filled with food additives, pesticides, and other chemicals." All you need is the breakthrough "100-percent natural pill" that "can put your doctor out of business." Is this really that different from Ching's Patent Worm Lozenges, advertised in 1802 as being of "peculiar importance to those afflicted with internal complaints"? It seems human credulity is a constant, unaffected by the march of science. Now do you see why I collect ducks?

The "Cancer Conspiracy"

"As a crab is furnished with claws on both sides of its body, so, in this disease, the veins which extend from the tumor represent with it a figure much like a crab." So wrote Roman physician Galen 2,000 years ago, speculating on why some 600 years earlier Hippocrates had used the Greek word "carcinos," meaning "crab," to describe abnormal growths on the body. Our word "cancer" is the Latin translation of "carcinos." Although doctors long ago learned to recognize this fearsome disease, they didn't have much to offer in terms of treatment. It certainly wasn't for lack of effort. Over the years, physicians tried everything from pulverized crab ointments to cauterizing cancerous lesions with red-hot metal. Some even resorted to "sympathetic magic," believing that placing a live crab on a tumor would allow the disease to be transferred to the animal. Such methods had about as much chance of success as the various cancer "cures" that populate the web today.

Being in the science communication business requires currency with both the scientific and pseudoscientific gusts of information that blow through the internet. That's why I subscribe to a large number of news feeds, including ones with seductive titles such as "Cancer Defeated," "Underground Health Reporter," "Step Outside the Box," "Natural Cures Not Medicine," "Nutrition and Healing," "The Alternative Daily," "Expression of Truth," and "Reality Health Check." Although these newsletters have various agendas, they do have a common theme: there is a conspiracy between "Big Pharma" and mainstream medicine to hide effective "natural" cancer cures from the public. Regulatory agencies are also seen as part of the conspiracy, accused of being in the pocket of multinational corporations who of course are out to destroy people's health.

Luckily, we are told, there are "maverick scientists" out there who "swim bravely against the tide to tell us about their scintillating breakthroughs that mainstream medicine refuses to embrace." There is talk of "insider secrets that stop cancer in its tracks" and promises of "exposure of mainstream medicine's deadliest conspiracies." Of course, "you can't hear about these secrets from your doctor, but you shouldn't blame him because his hands are tied, and he could even lose his license for recommending safe, natural alternatives to toxic cancer drugs." Hogwash!

Often the newsletters feature a video that we are urged to view quickly because "it might not be available for long since powerful interests are hell bent on minimizing the damage it is doing to corporate medicine's profit machine." Gee, aren't we fortunate to have all these daring doctors and scientists who are willing to reveal their scintillating, cutting-edge, dazzling research as they "battle vested interests' liars-for-hire who campaign to discredit the shocking truth."

Make no mistake about it, all these champions of "alternative" treatments have their own vested interests. There is always a book to buy, a "health letter" to subscribe to, or a product to purchase. Often the hook is a video that describes some gallant researcher whose natural cure for cancer was laughed at, but which, according to the testimonials provided, "produces such spectacular results that the only side effect is chronic good health." There are all sorts of allusions to the wondrous treatment, but the actual "cure" is not revealed — at least, not until you sign up for a subscription. Well, I've signed up for a good number and I have learned, for example, how "one courageous MD, who spent his career proving that nobody does it better than Mother Nature," will reveal, for a price, a cancer treatment that has a "100 percent success rate backed by 80,000 cases." What is it? Turns out to be eggplant extract! I don't know where all those successful cases are, but they certainly are not recorded in the medical literature. There are a couple of reports of eggplant extract having some efficacy on basal cell skin cancer in a few patients. Hardly a magnificent cancer cure!

Another of my newsletters offers to stop cancer in its tracks with the "Fruit of the Angels." It turns out to be papaya. As is usually the case, there is a seed of scientific fact that the author nurtures into an orchard of folly. Some papaya extracts have been shown to slow the multiplication of cancer cells in laboratory cultures, which is really a ho-hum observation. Numerous substances do this with little clinical relevance.

Yet another of my sources claims that the "King of Plants," so dubbed by the Chinese, is the answer to cancer. The king happens to be the chaga mushroom. There are references to antioxidant properties, as well as to Nobel laureate Alexander Solzhenitsyn's classic book, *Cancer Ward*, a semi-autobiographical novel in which a character cures himself of cancer with the mushroom. The Soviets apparently embraced the treatment but somehow managed to keep this crowning achievement from the West. Sure.

"These people don't get cancer until they move away from their native land and change their diet," squeals yet another newsletter. It goes on to say, "a century ago a British doctor stumbled across an isolated tribe in India where cancer was unknown." I had to purchase the book that was being promoted to find out that their secret was a diet high in apricot pits! It's reminiscent of the patented drug Laetrile, a totally debunked cancer treatment that the book promotes with religious fervor.

And how about "thunder god vine," "the true cancer killer that stunned scientists by wiping out cancer in forty days," according to yet another bulletin. Well, not quite. Researchers actually found some efficacy with a synthetic analog of triptolide, a compound found in the vine — in mice. The author of this epic goes on to take issue with researchers trying to create a pharmaceutical drug that can be patented, and counsels people to just get their hands on natural thunder god vine. Nonsense. It doesn't work. That's why the synthetic analog was tested.

There are also numerous newsletters that promote a

variety of superfoods with remarkable health-enhancing and life-extending properties. Such as a magical mix created by Mister B, a Beverly Hills millionaire who decided he didn't want to age. After five years of research, he distilled his list of superfoods down to chlorella, moringa, maca, spirulina, cacao, wheatgrass, camu-camu, and acai. Of course, you don't have to go searching for these; they are all available in one jar.

Why go to all this trouble, though? Why not just drink lemon juice, which, according to a widely circulating email, miraculously kills cancer cells and is 10,000 times stronger than chemotherapy? Scientifically bankrupt slop. But you can take this to the bank: there is no conspiracy to keep cancer cures from the public! If you do want to look for a conspiracy, take a look at those who are trying to make a buck from promoting the idea that such a conspiracy exists.

Yikes, I'm Infested!

I have little insects inside me that are dining on my cartilage, bones, and muscles. It seems they invaded my body either from animals or from dirt. These bugs used to eat plants, I'm told, but because they've been genetically modified, they now eat us. I also have a type of worm in my blood vessels. These creatures come in couples with the female living in the male body. I also have an overabundance of vitamin C in my kidneys and an inflammation of the sciatic nerve caused by a plasma virus. My prostate gland is infected by a brown mushroom. My red blood cells are a little too big due to microbacilli that are either released by plants in my office or come from eating fruits that weren't washed properly. Apparently these bacteria like to eat the fat from the red blood cells, which then causes them to become bigger. (The blood cells, not the bacteria.) I also have a viral infection in my right eye. And my muscles don't work properly because mushrooms have grown roots that tangle the muscle strings. I guess it's a wonder that I'm still alive.

I'm not too worried, though. The worms, bacteria, mushrooms, and viruses were not revealed by blood tests or CAT scans. They were diagnosed by a different kind of scan. I was informed of all the nasty action going on inside my body by a clairvoyant/naturopath who scanned me from top to bottom with her eyes closed, sensing, as she claimed, "life frequencies." Needless to say, my problems were "treatable."

Let's rewind a little. This little adventure started with an email I received that intriguingly began with: "In the past, those like me were called witch, saint, gifted, mutant, freak, and more . . . but I have an extraordinary ability at being able to find elements and microscopic life such as bacteria, viruses, worms, parasites, and algae in the human body, the earth's crust, and so on." The writer assured me that this was not a hoax and was looking to be tested in exchange for a document attesting to her ability. I was game and we discussed various ways that her abilities could be put to a test.

She told me that "when looking through a human, I see chlorine as yellow bubbles; radon as a pale blue accumulation; copper as white." These claims really weren't testable but we hit on something when she mentioned she could see germs in water and could distinguish between tap water, bottled water, and lake water. We settled on a challenge that involved randomly placing one of these waters into each of fifteen glasses. Her task was to identify the samples. She actually got eight correct, but that fell short of the ten that we had agreed would constitute a meaningful result. I asked if water that had no germs would be easier to identify and she thought that would be the case. So I set up four glasses that contained either tap or distilled water. She only got one of these right.

I thought we were now done with the experiment, but was told that actually her main talent was diagnosing what was going on inside the body and she was quite willing to demonstrate this ability. And so we began. Her very first words were "this is for entertainment purposes only," which was fine with me, as I did think this would be quite entertaining. "There's a lot of carbon in your system, especially in the liver and the blood." Well, she got that right. All the proteins, fats, carbohydrates, and nucleic acids that make up our tissues are organic compounds, meaning their basic structure is built of carbon atoms. I don't think, however, that is what she had in mind.

Next, I was told I have a lot of heavy metals in my lungs, like machinists who solder a lot. I think I soldered once in my life. I must have smoked in the past, she went on, because I have a lot of "old" carbon in my lungs. I have never smoked. She also diagnosed schistosomiasis, a parasitic disease that causes my legs to be itchy. Schistosomiasis is an infection widely seen in Africa and Asia, never in North America. And my legs do not itch.

I have a purplish color in my liver. I was told that what you eat dyes your body, and I must have been eating beets. Nope. Can't remember the last time I ate this vegetable. Then I was told that I often get pain in my rib area from coughing or from rotating movements but I should not worry because a chiropractor can easily fix that. I have no such pain, and should I encounter it, my choice of treatment would surely not be a chiro. After scanning me, she did the same with two colleagues who were also filled with mites, insects, "phages," "microplasm infections," and who knows what else. In one case, she even claimed to see a tumor, specifically in the left testicle.

While performing these scans, she also revealed that she was able to communicate with the dead, and she volunteered to do readings for the three of us. It was amazing! Why? Basically, because she got nothing right! My grandparents made no mention of the fact that they died in the gas chamber, and my father must have been vigorously exercising on the other side because I was told he was a large muscular man. Actually, he was shorter and smaller than me. I was also told that the reason I'm constantly searching for my keys is that the mischievous spirit of a girlfriend I left for my wife was hiding them. Nope and nope. No such girlfriend, and I don't lose my keys.

Up to this point, I had been sitting straight-faced without making any comment because I'm quite familiar with "cold reading," and the ability of "psychics" to capitalize on any reaction from their subject. But now I suggested we discuss the happenings and explained that she had been off-track on virtually everything. At this point, she became agitated and asked why we had invited her if we were just going to waste her time, forgetting that she had sought the invitation. In any case, the clairvoyant then got up and muttered something about the failure being due to my skepticism that blocked her abilities, apparently not having foreseen this possibility. We never did get around to treatments, which I suspect were of the herbal variety. Next time, she said, she would seek out a microbiologist with an open mind and prove herself.

Anyone with a scientific background would of course recognize the garbled rhetoric we heard as total nonsense, albeit somewhat entertaining. But it was also clear that this clairvoyant/naturopath has clients who accept her abilities as more than just fun. And that isn't funny.

Full of It

I'm told I'm full of . . . ummm . . . "crap." And so are most of you. Literally. So say the promoters of various colon cleansers. Actually, they're not quite so crass; they prefer to use gentler terms, such as fecal matter, impacted waste, or spackle. But the message is clear: our colons are loaded with a repulsive noxious sludge, the result of an improper diet and a "toxic" environment. This putrid goo sticks to the wall of the colon, boosting our body weight. Even worse, it releases its foul contents into our blood, poisoning our entire system. The result? A nation of bloated sickies who lack energy and mental clarity. The unnamed toxins are powerful indeed, causing, we are told, ailments ranging from asthma, allergies, and prostate problems to cancer, heart disease, and an impaired sex drive. But luckily, there is salvation in sight. We can sweep the fetid guck out of our colon with one of myriad colon cleansers that compete for our attention and, of course, for our dollars, via ads that populate radio waves, magazines, and the internet.

And what spirited and imaginative ads! One product claims that we have anywhere from 6 to 40 pounds of waste, feces, and undigested food stuck in our bodies. Another one compares the weight of the waste to carrying a bowling ball in our gut. Then there are accounts of famous people who died and were found to be full of intestinal sludge. John Wayne, depending on which product's info you're reading, was found to have anywhere from 40 to 80 pounds of impacted matter in his colon. A curious claim, given that no autopsy was performed on the Duke. But the most inspired ads are the ones that provide us with a visual extravaganza of the "mucoid plaque" that is eliminated by users of colon cleansers. The pictures show the relieved patient holding the cause of his former misery, a long, gummy-looking, repulsive excretion. This, we're told, is the toxic guck that had built up in his colon over years before making a triumphant exit, stimulated by the wondrous colon cleanser!

Now, let's get real here. Have pathologists who have carried out thousands of autopsies seen pounds of goo encrusted in intestines? No. Have colorectal surgeons who have operated on colons thousands of times seen such sludge? No. Have radiologists who have perused thousands of X-rays of the colon noted the buildup of "mucoid plaque"? No. Why? Because it doesn't exist. The term itself was the invention of naturopath Richard Anderson, who created Arise and Shine, a popular colon cleanser. So what, then, is the yucky stuff that has emerged from the rear of a happy colon cleanser devotee that we see revoltingly displayed in those photos on the web? Supposing that the pictures are not faked, I suspect what we are looking at is the colon cleanser itself making an impressive appearance.

Although the specific ingredients in these products vary, they all contain some sort of laxative, be it a fiber blend or an extract of cascara sagrada bark, well known to stimulate intestinal contractions. Classic fibers include psyllium husk, flaxseed, fennel seed, slippery elm bark, apple pectin, and guar gum. All of these can send you running in a hurry. And they are indeed prescribed for that very purpose by physicians. But problems can arise. Fiber absorbs water in the gut and sometimes can swell, making it difficult to expel. Usually this is prevented by drinking lots of water, which helps flush out the fiber before it has a chance to expand and form an intestinal blockage. In rare cases, with just the right (actually wrong) amount of water consumed, the mixture of fibers can be expelled as a long, stringy, slimy glop. The likelihood of this happening is increased if the colon cleanser contains bentonite clay, sometimes included for its ability to "absorb toxins." Such an impressive excremental display would be very rare, and certainly not something that all users should expect, contrary to what the promoters imply. And most assuredly the disgusting exudate is not any sort of toxic buildup being expelled.

Of course, just because the pounds of intestinal gunk only exist in the sluggish mind of some quack, we can't assume that products that help to evacuate the colon more regularly have no merit. What we need, though, are not baseless statements like "a dirty colon is a breeding ground for disease," or testimonials from users about how their bad breath, dizziness, irritability, or "brain fog" were resolved after scrubbing and buffing their colon. How about some evidence?

Well, you can search the scientific literature high and low and you will not find any proper controlled trial of colon cleansers showing they have any health benefit. How about problems? Possible. Back in the early 1990s, guar gum, an ingredient present in some colonics, was banned from diet products in the U.S. At the time, chewing gum with added guar gum was a hot seller because it was supposed to curb the appetite by filling the stomach as it absorbed water. It did, but it also caused esophageal and intestinal blockages. And yet, there it is today in some colon cleansers. One of these actually makes the claim of weight loss as it uses guar gum to "evercleanse" the pounds and pounds of (nonexistent) "spackle" from the colon. It is not the colon but the absurd claim that needs to be cleansed.

While the cleaning effect of colonics on colons is questionable, their effect on cleaning out bank accounts is not. A month's supply needed to "dredge toxic sediment" can run up to a tidy little sum. Why not spend the money on what goes into the colon, rather than on what comes out of it? A diet high in whole grains, fresh fruits, and vegetables is what your colon and the rest of your body really needs. Granted, your output may not be quite as spectacular as the samples seen in those colon cleanser ads, but you and your bank account will be healthier. Of course, if you are a fervent believer in colon cleansing, you will not be deterred by my arguments and will remain convinced that, unlike you, I'm full of crap.

Poking into the Pukeweed Doctor

"I think we never had more need to be on our guard than at the present time. The people are crammed with poison drugs and the laws say they shall not examine and judge for themselves. The effects are pains, lingering sickness, and death. Poison given to the sick by a person of the greatest skill will have exactly the same effect as it would if given by a fool."

You might think that quote comes from one of the numerous current websites that espouse the benefits of "natural treatments" over pharmaceutical drugs. It doesn't. It was actually uttered some 200 years ago by Samuel Thomson, an uneducated pig farmer whose philosophy that any man could be his own physician took America by storm in the nineteenth century. Thomson's model for self-help healthcare was eventually embraced by more than 3 million Americans, and his ideas even spread to Europe!

Thomson believed that all diseases could be cured by the use of herbs and heat. While his system of healing used some sixty herbs, *Lobelia inflata*, also known as pukeweed or Indian tobacco, was front and center. Pukeweed is a very appropriate name because ingesting the flowers, seeds, or roots of the plant makes people, let us just say, lose their breakfast. Thomson believed that before healing could commence, toxins had to be eliminated, and pukeweed was just right for the job. This was not a novel idea; conventional physicians at the time used mercurous chloride, better known as calomel, to purge patients. Lobelia's effects were less violent, and Thomson's theory that people could cure themselves without relying on doctors appealed to a lot of people. Thomson was not the first to experiment with lobelia. Native Americans treated dozens of ailments with the herb, ranging from fevers and venereal diseases to earaches and stiff necks. Lobelia also had a reputation as a love potion, which is hard to explain. Vomiting and love usually don't go together.

In Thomson's regimen, after the pukeweed had finished its performance, it was time to restore the body's heat with steam baths and cayenne pepper often in the form of an enema. If there were still complaints, other courses of treatment would follow, and complex mixtures of herbs such as ginseng, peppermint, and horseradish were often mixed with camphor and turpentine.

As is often the case for "alternative therapies," Thomsonism was rooted in its patriarch's personal experience. Young Sam had become curious about a plant that grew wildly in his father's fields and for some strange reason, tried chewing its pods. The effect was dramatic. It seems the man whom skeptics would eventually call the "puke doctor" had a funny bone. He convinced some of his friends to sample lobelia and had a good laugh at their expense. His interest in plants aroused, Thomson began to follow the healing abilities of an "old wife" in the area who had a reputation for curing people with herbs, often consumed as a brew in hot water to produce sweating. He was intrigued when she managed to cure his rash with a concoction of herbs. And then came a couple of catalytic events.

At the age of nineteen, Thomson sustained an ankle injury that defied conventional treatment but resolved when he ingested comfrey root and applied a turpentine plaster. Two years later, his mother contracted measles, which turned into what doctors called "galloping consumption." Thomson later commented that this had been an appropriate name because the doctors were riders who managed to gallop her out of the world in about nine weeks. But when he contracted the disease, he claimed to have cured himself with herbs. When Thomson later saw his wife cured by herbalists after doctors had failed, and he himself managed to cure his infant daughter of some skin condition by holding her over steaming water, Thomsonism was ready to gallop. Doctors, or "educated quacks" as Thomson called them, may have had their fancy degrees, but their blistering, bleeding, and purging were worse than useless. He could cure people with herbs and steam! Herbs grew toward the sun, the life-giving source of heat, and therefore must refresh one's health, the puke doc maintained.

As one might expect, physicians didn't take kindly to Thomson's attacks, and in 1809, one actually managed to accuse him of killing a patient with an overdose of lobelia. The puke doctor had to await his trial in a cell for six weeks. At the trial, he claimed that he had actually cured the patient, who was responsible for his own demise by venturing out into the cold instead of recuperating in a warm house. Meanwhile, the prosecution claimed that the victim had succumbed because of excessive vomiting brought on by lobelia. It is unlikely that this was the case, because lobelia does not induce such dangerous vomiting, but Thomson was exonerated because of a botanical error by the prosecution. An astute defense attorney noted that the plant the prosecution had introduced as evidence was actually marsh rosemary and not lobelia. That was enough for the case to be dismissed.

Thomsonians regarded the dismissal as vindication of their

efforts and the movement continued to pick up steam. Indeed, it was the popularity of Thomsonism that led to the repeal of the laws that a number of states had passed restricting the practice of unconventional medicine. Opponents had labeled these "Black Laws" in reference to those that restricted black Americans from practicing medicine. Eventually, Thomson's movement faded when some of his followers grew tired of his attacks on physicians and his drive to end physician licensing. They wanted more legitimacy and urged more training and even the establishment of Thomsonian hospitals. That never happened, but Thomsonism holds a unique place in history as a pivotal factor in allowing unconventional treatments to legally flourish in spite of a lack of evidence for efficacy. It is one of the pillars upon which modern naturopathy rests.

Vinegar Claims Leave a Sweet and Sour Taste

In 218 B.C., the Carthaginian general Hannibal crossed the Alps with his elephants to settle a score with Rome. The perilous journey almost came to an end when his army approached what looked like an impenetrable rockfall. But Hannibal, an ingenious leader, had a trick up his sleeve. Or, at least, he had some vinegar in his pot. As the Roman historian Livy recounts, the general had his men heat up the vinegar and pour it over the rocks, causing them to crumble. And here the story crumbles. Scale deposits in a kettle may certainly crumble when immersed in hot vinegar, but that is a long way from breaking down a wall of rock, even if it is made of limestone. Like scale in a kettle, limestone is made of calcium carbonate, which will react with the acetic acid in vinegar to form soluble calcium acetate and carbon dioxide. But there's no way that pouring vinegar on boulders will do anything but cause a bit of bubbling on the surface as carbon dioxide is released. So Livy's story has to be swallowed with a very large grain of sodium chloride, especially given that his account was written some 200 years after the supposed event.

This is not the only apocryphal story about vinegar. Here's a classic: during one of Europe's many plagues, four thieves in France had made a career out of robbing the dead. They were finally caught, surprisingly never having contracted the disease. In exchange for leniency, the thieves agreed to reveal their secret formula for avoiding the plague. It seems they had been drinking a concoction made by macerating garlic along with some other herbs in vinegar made from wine or cider. Even doctors bought into the tale. Whereas today the stethoscope is the symbol of the physician, back in the seventeenth century, it was a gold-headed cane with a hollow head filled with the vinegar that supposedly had kept the plunderers of the dead safe from the plague. Accounts of physicians sniffing the vinegar as they attended to the sick gave birth to the legend of the therapeutic properties of Four Thieves Vinegar.

Now fast forward to modern times. Four Thieves Vinegar, or "Vinaigre des quatre voleurs," is still with us. Its French producer claims to use a recipe identical to the "historical" version brewed in the sixteenth century. Well, that is as historical as Hannibal's rock-dissolving vinegar. In any case, the claim is that the concoction stimulates the immune system and offers better protection against the flu than a vaccine! But there's more: you can even apply it as a compress for treatment of arthritic pain and headache. Or use it as a rinse after shampooing to reduce frizziness. That actually does work. Of course, so does any old vinegar.

You don't have to travel to France to look for claims of all the marvelous things that vinegar, particularly the apple cider variety, can do. There are plenty of books, pamphlets, and ads promoting apple cider vinegar on this side of the big pond too. It's a "nutritional powerhouse" that fights cancer, curbs arthritis, reduces blood pressure, dissolves fat, cleans out "bad" cholesterol, reduces fatigue, treats ulcers, and even improves memory. Sometimes, though, regulatory authorities get fed up with the unsubstantiated blather.

"Jogging in a Jug," a dietary supplement with apple cider vinegar as a key ingredient, was created by former Alabama dairy farmer Jack McWilliams in the early 1990s with claims of providing the same health benefits as jogging, including alleviating heart disease and arthritis while "cleansing the internal organs." The Food and Drug Administration and the Federal Trade Commission determined these claims were unsubstantiated, and in 1995, penalized McWilliams's company, Third Option Laboratories, to the tune of $480,000. Furthermore, any future advertising had to state, "there is no scientific evidence that Jogging in a Jug provides any health benefits."

The company is still around today, run by McWilliams's grandson. No direct health claims are made, but the company's website features testimonials alleging effective treatment of allergies, chronic fatigue, high blood pressure, acne, and obesity. Abiding by the FDA ruling, there is the disclaimer that Jogging in a Jug has conducted no scientific research and that comments from customers have not been verified scientifically. It is clear, however, that people are not buying Jogging in a Jug to sprinkle on their French fries. Danny McWilliams Jr. certainly thinks highly of his company's product: "We believe this is the best dietary supplement ever made. One could say it is nature's own dietary supplement." Yes, one could say that. But that doesn't make it true.

There are numerous other promoters of various apple cider

vinegar products as well. Often they will infer that they are being prevented from making health claims because of the influence that pharmaceutical companies have with regulatory agencies. Why would anyone want to buy expensive medications when apple cider vinegar will do the trick, they ask? Maybe because the medications work and the vinegar does not.

So, is there any actual evidence that apple cider vinegar can provide any sort of health benefit? Perhaps surprisingly, there is. But the effect is far from earth-shaking. Dr. Carol Johnston at Arizona State University has shown that a couple of teaspoons a day may help improve blood sugar control in type 2 diabetics. It seems acetic acid inhibits some of the enzymes that digest sugar and starches, meaning that these are more likely to pass through the digestive tract without being absorbed and therefore have less of an impact on blood sugar.

What about the much-ballyhooed claim that apple cider vinegar will "melt the fat away"? Dr. Tomoo Kondo and his group at the Central Research Institute in Japan have looked into this. In a properly controlled double-blind study of 155 obese patients, they found that about four teaspoons of vinegar a day over three months resulted in a weight loss of about a kilogram, and a reduction in waist size of about 1.5 centimeters. Maintaining these losses, however, required continuous ingestion of vinegar. A possible explanation is that acetic acid interferes with some of the enzymes involved in lipogenesis, the conversion of sugars to fat. Interesting, but it's not of great clinical relevance. At least, though, vinegar doesn't cause any harm. Unless you spill it on your marble countertop, that is. Still, I think I'll continue to do my jogging on a treadmill and reserve the jug for other forms of nourishment.

Diagnosing Pathological Science

Back in 1953, Nobel Prize-winning chemist Irving Langmuir coined the expression "pathological science" to describe a process by which a scientist seems to follow the scientific method but unconsciously strays in favor of wishful thinking. Pathological science is distinct from fraud in that there is no intent to deceive. It is essentially faulty science promoted by people who are somehow blind to the evidence against their pet ideas. The most frequent use of the term over the last couple of decades has been in connection with "cold fusion," a phenomenon first proposed in 1989 by electrochemists Drs. Stanley Pons and Martin Fleischmann.

The two highly regarded researchers stunned the scientific community by calling a press conference to announce that they had detected the fusion of deuterium nuclei under simple laboratory conditions. They claimed to have evidence of release of energy that could not be explained otherwise. A wave of excitement spread around the globe with optimistic musings about the process that could be the answer to our energy problems. The euphoria quickly waned when other research groups were unable to reproduce the experiment. A general consensus soon emerged that Pons and Fleischmann had noted some anomalous phenomenon, but had misinterpreted their findings. It was wishful thinking, not experimental evidence, that had produced cold fusion. Actually, while the names of Pons and Fleischmann are most commonly associated with cold fusion, they were not the first to claim that such a process can occur under mild conditions.

Some thirty years before Pons and Fleischmann's press conference, French chemist Louis Kervran introduced his theory of "biological transmutation," claiming that in living systems, atoms of one element can combine with those of another to give rise to a third element. Potassium, Kervran suggested, can under the right conditions combine with hydrogen to form calcium. This was a stunning claim, an apparent realization of the classic alchemical quest to transmute one element into another. Could this be?

The identity of an element is determined by the number of protons in its nucleus. Potassium has 19 protons and hydrogen has 1. If in the body these somehow combined to form one nucleus of 20 protons, we would indeed have calcium. We would also have an event that defies everything we know about chemistry and physics. Elements don't combine to form new elements except in the case of nuclear fusion reactions, which require a tremendous input of energy, only achievable at temperatures of millions of degrees. Such conditions are met in our sun, where hydrogen nuclei combine to form helium, accompanied by the release of vast amounts of energy. But outside of such extreme conditions, chemical reactions cannot create or destroy atoms, they can only rearrange them to form novel molecules.

Kervran had originally been intrigued by a question raised by French chemist Louis Nicolas Vauquelin in the late 1800s: how could hens manage to produce eggshells, which were composed of calcium carbonate, in spite of being fed a diet of oats, known to be very low in calcium? Kervran concluded that potassium, which is plentiful in oats, must combine with hydrogen to produce calcium. He dismissed the problem of the immense energy requirement for such a process by suggesting the existence of a "low-energy transmutation." This would become known as the "Kervran effect." Kervran also claimed that a crayfish placed in a basin of seawater with all calcium removed still managed to make a shell. Again, he suggested, potassium must have been converted into calcium.

According to Kervran, even the strange case of industrial carbon monoxide poisoning when no carbon monoxide had actually been inhaled was to be explained through transmutation. Nitrogen gas, which makes up about 80 percent of air, is composed of two nitrogen atoms joined together. Each nitrogen atom has seven protons in its nucleus, and if a proton from one nitrogen atom were to be transferred to the other, two novel nuclei, one with six and the other with eight protons, would be formed. In other words, we would now have an atom of carbon and one of oxygen, thereby explaining the formation of carbon monoxide. But chemistry doesn't work by such simple arithmetic. Such a transfer of protons from one nucleus to another has never been observed.

In spite of the scientific implausibility of biological transmutation, Kervran's theory was not dismissed out of hand by all. Italian researchers carried out a carefully controlled study of oats under a variety of conditions, analyzing for calcium, potassium, and magnesium. There was no evidence of any kind of transmutation. What, then, about the egg-laying conundrum? Actually, there is no conundrum. If there isn't enough calcium in the chickens' diet, the birds will mobilize calcium from their bones. Furthermore, oats are not devoid of calcium; the early analyses back in the nineteenth century were faulty. Very simply, the calcium needed for eggshell formation must somehow be provided in the diet. It doesn't come from any sort of transmutation. In modern egg-laying facilities, the diet of the hens is often supplemented with crushed oyster shells, cuttlefish, or crushed limestone to ensure adequate calcium intake. Sometimes even eggshells themselves are recycled in feed.

Kervran's idea about carbon monoxide production in the body from nitrogen was also wrong. There is no mystery about finding carbon monoxide in the blood of people who never inhaled any. It forms naturally in the body when an enzyme called heme oxygenase reacts with heme, a breakdown product of hemoglobin. The bottom line is that the Kervran effect doesn't exist. The French chemist simply came to the wrong conclusion based on some faulty observations.

It is curious that, in spite of being a competent and respected scientist, indeed an expert on radiation poisoning, he was willing to propose a theory that flew in the face of established science. No wonder he was awarded the 1993 Ig Nobel Prize for physics in recognition of his conclusion that calcium in chicken eggs can be created by some sort of biological transmutation. Were such a fusion process to occur in a chicken, the energy released would turn the bird into an atom bomb.

The Ig Nobels are awarded annually to "honor achievements that first make people laugh and then make them think." Kervran's "biological transmutation" was well worthy of the award. If there were awards for pathological science, Kervran would be a high-ranking candidate. Of course, there would be many others jostling for spots on that list of candidates.

Mountebanks

I would like to propose that the word "mountebank" be updated to "mounteweb." Let me explain. The original term derives from the Italian "*monta in banco*," which literally means "getting up on a bench." So, mountebanks were sellers of dubious medicines who would "mount" on a "bench" and regale a gathering crowd with descriptions of their wondrous nostrums and elixirs that promised to restore health and endow men with unparalleled sexual powers. By the fifteenth century, mountebanks were to be found on many a street corner in

Europe, often accompanied by a "Merry Andrew," whose task was to attract an audience with an assortment of zany antics.

Historians suggest that the original Merry Andrew was actually Dr. Andrew Borde, physician to King Henry VIII, who was noted for his wit and captivating way of addressing the public on health matters. He produced merriment in his audiences and gave rise to imitators who may have lacked his knowledge but nevertheless managed to entertain the crowds with their buffoonery. These clowns came to be known as Merry Andrews and by the seventeenth and eighteenth centuries, no mountebank would be without his Merry Andrew.

The idea of blending comedy with medicine has historic origins. The famous anatomical theater at Bologna, where dissections were performed for medical students as early as the sixteenth century, featured a small door just above the lecturer's platform. When the professor found the students to be inattentive, he gave a signal, the door opened, and a fool's head would pop through. He cracked a joke and quickly withdrew. Students were roused from their somnolence, had a hearty laugh, and refocused their attention on the lecturer's words.

Once the Merry Andrew had attracted a crowd, the mountebank would do his best to liberate coins from pockets and purses. One of the most notorious quacks was Ben Willmore, whose spiel was actually recorded by an onlooker. "Behold this little vial, which contains in its narrow bounds what the whole universe cannot purchase, if sold to its true value. This admirable, this miraculous elixir, drawn from the hearts of Mandrakes, Phoenix livers, Tongues of Mermaids, and distilled by contracted Sunbeams, has, besides the unknown virtue of curing all distempers both of mind and body, that divine one of animating the Heart of man to that degree, that however remiss, cold and cowardly by Nature, he shall become Vigorous and

Brave. Gentlemen, if any of you present was at Death's Door, here's this, my Divine Elixir, will give you Life again." Wow!

Some mountebanks attacked physicians, much as is the case today. One Tom Jones was a classic example. He would rail against doctors whose only remedy for disease was to purge or bleed the patient. Of course, he had the real solution. His "Incomparable Balsam" healed all sores, cuts, and ulcers, his "Specifick" cured pain in a minute, and his "Pulvis Catharticus" expelled poisons and fortified the heart against faintness. Actually, while probably useless, these nostrums were less likely to harm the patient than doctors' bleeding or purging.

Some of the mountebanks were more audacious in their challenge to physicians. John Pontaeus gained fame in the seventeenth century with "Orvietan," his antidote to all poisons. He even offered proof. Physicians could administer a poison of their selection to his assistant, who would then be treated with a dose of Orvietan. The doctors accepted the challenge and decided on Aqua Fortis or, as we now know it, nitric acid. Not only was this known to be toxic, it was also highly corrosive. The quack's servant swallowed it, collapsed immediately and was carried away, apparently dead. To the surprise of the physicians, he reappeared the next day, none the worse for wear. It seems Pontaeus had a trick up his sleeve. Or, more accurately, butter down his servant's throat. Before the "experiment," the assistant had swallowed a large dose of butter, enough to coat his mouth and throat, protecting him from the caustic liquid. After being carried off, he was immediately given warm water and the water-butter mix made him so sick that he regurgitated the acid. So the story goes.

Pontaeus pulled other fast ones. He sold a "Green Salve" that supposedly healed all wounds, and he had an impressive demonstration to prove it. His assistant dipped his hands into molten lead after which the apparently badly burned hands would be restored to perfect health with the magical salve. The audience happily anted up for the wonder product. I hope they didn't try to reproduce the molten lead experiment because Pontaeus's "molten lead" was actually mercury dispensed with a ladle painted red to give the appearance of heat. The assistant's bloody hands, which were displayed to the onlookers after being withdrawn from the "lead," were actually colored with vermilion (mercury sulphide) that had been hidden in his hand as he dipped them into the "molten" metal. The spectators were properly duped and Pontaeus's pocketbook swelled. As far as the assistant went, his occupational hazard was mercury poisoning.

There were other ingenious performances as well. In the early seventeenth century, an Italian mountebank became famous for having healed his arm with a miraculous oil after he had just gashed it with a knife. The healing oil was effective indeed, as long as the mountebank was equipped with a trick knife and had mastered the art of palming a piece of fabric soaked in chicken blood. Unfortunately, sometimes such effects backfired, with the performer being accused of witchcraft. A young mountebank in Cologne was charged with witchcraft for having torn and restored a handkerchief in the presence of witnesses. What happened to the unfortunate performer is not known, but the magic trick has certainly survived. It's one of my favorites.

Today, mountebanks have transformed themselves into what I propose to call "mountewebs." Instead of beguiling a few onlookers by mounting a bench, they snare multitudes by mounting websites. But their "structured water," "detox foot baths," "ear candles," and "energy bracelets" are no more effective than Willmore's phoenix livers, mermaid tongues, or contracted sunbeams.

Double Helix Water

Unfortunately, chemistry is a mystery to many. And that suits the hucksters just fine. It sets the stage for cashing in on chemical ignorance by bamboozling people with scientific-sounding balderdash. Ignorance, though, is not total. There is one molecular formula that people do tend to recognize, and that is good old H_2O. Then if you press them to name an important chemical in the body, chances are they will come up with DNA. And they are likely to have some sort of mental picture of the double helix structure of DNA, since after all, it's been widely featured in popular books, movies, and TV shows. You can hardly miss the huge model of DNA on the set of *The Big Bang Theory*!

Given that both water and DNA are generally recognized as essential to life, it comes as little surprise that bottles of "Double Helix Water" have appeared on the scene. The label lists "pure water" as the only ingredient, but does feature a reference to a publication in a physics journal about "stable water clusters," followed by the disclaimer that "the company does not endorse claims or have scientific proof that stable water clusters are effective in the cure, mitigation, treatment, or prevention of disease." Obviously, though, the intent is to infer health claims. Why else would the Double Helix Water website feature testimonials about improved energy, reduced pain, better sleep, and improved mental clarity? And why would a book by the dynamic duo that promotes this water feature on its cover the question "Could this discovery save your life?" And why would that book have a series of pictures that purport to be infrared images of cancer patients before and after drinking Double Helix Water? I have no idea where those pictures actually come from, but the implication is that this preposterous product has some sort of effect on cancer.

Whatever the power of Double Helix Water is, it must be potent. Why? Because the water has to be diluted to be used. Imagine the nonsense of diluting water with water. According to the instructions, you just add three to four drops to a glass of distilled water and then drink two glasses a day. The mumbo jumbo that explains how Double Helix Water is supposed to work is astounding. Here's a gem: "Stable Water Clusters found in Double Helix Water may act as the body's fundamental building block; a foundation for self healing and protection from environmental toxins. The water may help bypass blocked Meridians and allow qi to flow again." And, of course, it may not. Meridians are mythical channels through which the mythical qi energy flows.

What about the notion that this "newly discovered phase of water can unravel the differences between allopathic and homeopathic medicine." Just what is this newly discovered phase of water? It's a figment of the promoters' imaginations. Water molecules do form associations with each other, with the partial positive charge on the hydrogen atom being attracted to the negative charge on the oxygen atom. At any given moment, these associations may be described as a cluster, but they only have a transitory existence, on the order of picoseconds, before the molecules rearrange to another cluster. These clusters have no observable properties and cannot be stabilized. Needless to say, they have nothing to do with any crackpot ideas of transporting toxins or building bridges over energy blockages. All that you get for forking over about sixty dollars for 15 milliliters of very ordinary water is a spectacular lesson in hucksterism.

It seems that it would be hard to outdo the Double Helix Water malarkey, but "TC Energy Design" gives it a valiant try. The real problem with water, you see, is that it is "weakened by flowing in straight pipes and by the unnatural high water pressure." That's why, before we drink it, it should be "energized and revitalized." "Consuming food and water of higher vibration supports you both consciously and unconsciously" and "vitalized water supports the purification processes of the body which is essential for health and well-being." And how do you get your water to vibrate properly? Simplicity itself. Just store it for three minutes in a carafe or glass created by Austrian composer Thomas Chochola, the "TC" of Energy Design. This is no ordinary glassware. Oh, no, it is balanced, harmonic glassware! Chochola has managed to "convert his musical compositions into spatial dimensions using mathematical calculations."

"The shape of the glassware," we are told, "generates an energizing resonance pattern that restores the water within and improves the surrounding environment with subtle waves of harmonic resonance." Needless to say, "all dimensions are musically fine-tuned with one another and with a 6-wave primary structure they emit a major triad, which can be mathematically expressed as a relationship of 1:3:5:8. These ratios can be observed in nature and stand in resonance to superordinate motion sequences in the cosmos. The engagement of the TC shapes with biological naturally occurring factors can be physically described as a coherence phenomenon." It can also be described as incoherent poppycock.

The TC website even has a page pompously titled "The Science." Here we find pictures of "water crystals" before and after the water is stored in the magical carafe. Never mind that there is no such thing as a water crystal. There are, of course, ice crystals, but no water crystals. The "biological valency" of the water also improves. This, I learned, is a measurement system "used by many dowsers and geomants to locate the vitality of humans and food." Yup, geomants. And what is a geomant? One who practices a method of divination by interpreting markings on the ground or the patterns formed by tossing handfuls of soil, rocks, or sand into the air. If you don't want to take the word of dowsers or geomants, how about the word of a Japanese laboratory that claims to have observed a decrease in stress levels thirty minutes after drinking TC water? I think I need to drink some of the stuff myself, because my stress level increases just by reading this nonsense.

Although the carafes designed by Thomas Chochola are pricey, a glass drinking straw created on the basis of "modern quantum physics as well as ancient insights into the natural flow of energy" is available for about twenty dollars. Of course, it "incorporates the principle of the vortex for an increase in energy level." Just imagine the astounding benefits of sucking Double Helix Water through this straw! Suckers are welcome to give it a shot.

Fakes, Phonies, and Impostors

Fake! Fake! Fake! Fake blueberries, fake cosmetics, fake fish, fake drugs, fake pesticides, fake science, fake experts. It is the Age of Fakery. Phony blueberries may not have a big impact on health, but counterfeit malaria drugs can have devastating consequences. And what's the motivation behind the extensive fakery? What else? Money.

In recent years, the antioxidant content of blueberries has stimulated the palate of researchers, leading to a variety of publications that have attracted the media spotlight. They've also attracted the interest of food producers who are quick to take advantage of the public's yen for "superfoods." But blueberries are expensive, while compacted bits of sugar, corn syrup, starch, hydrogenated oil, artificial flavor, and artificial coloring are cheap. So that's what you'll find masquerading as real berries in some cereals, bagels, and muffins. That's misleading enough, but General Mills' Total Blueberry Pomegranate cereal takes deception to new heights. It contains no blueberries or pomegranate at all. Justification is to be found in the small print that informs us of the presence of "natural and artificial flavors." I think I'll stick to my steel-cut oats and add real blueberries.

When it comes to cosmetics, fakery takes on a different form. Counterfeit products that look like regular consumer items made by well-known companies are hitting the shelves. Makeup, perfumes, and shampoos may sport packaging that is identical to the original, but the contents can be significantly different. Tests of seized products, almost always originating in China, have shown higher-than-acceptable levels of metals that are known to cause allergic reactions. Copycat shampoos with name-brand labels have been found to harbor high levels of bacteria. Legitimate manufacturers are understandably up in arms over the proliferation of counterfeit imports, because any customer who buys a sub-standard product, believing it to be real, is unlikely to try the genuine one again.

Fish fakery has many dimensions. Farmed salmon are often passed off as wild, and supposedly local catfish may actually be *Pangasius hypophthalmus*, or "sutchi" catfish, imported from Vietnam, where some fish farmers use drugs that are unapproved in North America. And then there are Chin Chin fish masquerading as Garra Rufa. This isn't an earth-shaking problem unless you are into fish pedicures. You can go to a "fish spa" and immerse your feet in a tub filled with fish that nibble on dead skin cells, supposedly aiding skin repair and regeneration. This is more than silly; it's scary. Although some spas use filter systems and ultraviolet light to kill bacteria in the water, there have been cases of people picking up nasty infections, most likely from some microbe left by a previous customer. The situation is made worse by using fake Garra Rufa, because the impostor Chin Chin can bite hard enough to penetrate the skin and allow easier transmission of disease.

However, it is the prospect of fake drugs that really raises concern. The global trade in counterfeit pharmaceuticals is estimated to exceed $75 billion a year. Phony versions of Lipitor and Viagra, with no active ingredient, are widely sold on the internet, and various herbal remedies for arthritis or weight loss have been found to contain undeclared prescription drugs that have been outlawed in the U.S. and Canada. But perhaps the most insidious example is the proliferation of fake malaria pills, which may be responsible for some 200,000 deaths a year. The criminals' target is artesunate, a semi-synthetic derivative of artemisinin, the anti-malarial compound isolated from the sweet wormwood plant to which mosquitoes are developing a resistance.

Since malaria is widespread in Asia and Africa, the profits from selling cheap, fake artesunate can be stunning. And the counterfeiters are very good. Guilin Pharmaceutical, the Chinese company that manufactures the authentic version, has tried to distinguish its product from copies by adding a hologram to the packaging. In at least one case, a fake hologram was better than the original. Some of the fakes actually contain small amounts of arteminisin to foil testing by authorities, a practice that is particularly reprehensible. The dose is too small to have an effect, but is enough to increase the risk that the malaria parasite will develop a resistance to the drug the same way it developed resistance to chloroquine, the previous standard treatment. Although the origin of the fake pills has been traced to an area in southern China through the identification of pollen in the pills that is unique to the region, the actual manufacturer has never been identified. The mass murderers making the fake drugs remain at large.

The counterfeit pesticide industry also amounts to a multi-billion-dollar operation. As much as one quarter of the pesticides used in Europe may be fakes, originating mostly in China. These may be just total rip-offs in the sense that they are packaged in legitimate-looking containers but contain no active ingredient, or may contain pesticides that are illegal in Europe, or may be formulated with toxic solvents. They are usually distributed by organized crime. Numerous farmers have lost total crops because of the fake chemicals, and their effect on the environment and human health remains unknown.

Fake science oozes on the internet. Princeton researchers supposedly found that mice fed food stored in a fridge festooned with refrigerator magnets developed cancer, while a control group of mice did not. No such research exists. Nor has any Johns Hopkins research shown that cancer is curbed in an alkaline environment, or that cancer cells cannot thrive in an oxygenated atmosphere.

And then there are the fake experts, way too numerous to list. But Gillian McKeith, a prominent "nutritionist" in England whose degree is from an unaccredited U.S. college, is an example. Let's leave the last word to her: "I always think of the tongue as being like a window to the organs. The extreme tip correlates to the heart, the bit slightly behind is the lungs. The right side shows what the gallbladder is up to and the left side the liver. The middle indicates the condition of your stomach and spleen, the back, the kidneys, intestines, and womb." I wonder what she would say a forked tongue indicates.

The Trouble With Homeopathy

Homeopathic products are safe enough, no doubt about that. Millions of people around the world swear by them. No doubt about that either. Furthermore, their label features the term "DIN-HM" (Drug Identification Number-Homeopathic), designating approval by Health Canada. So why, then, do I and my colleagues at the McGill Office for Science and Society support a class-action lawsuit launched against Boiron Laboratories and Shoppers Drug Mart for marketing Oscillococcinum, a homeopathic medication advertised as a remedy for colds and the flu?

I have absolutely no desire to limit anyone's freedom of choice when it comes to selecting healthcare products, or any company's right to sell items that the public wants to buy, as long as the items are safe. But I do have a desire to ensure that whatever choice consumers make is based on scientifically informed opinion. In the case of homeopathy, misinformation can have consequences ranging from a needless waste of money to an individual foregoing more effective treatments. As an educator, I am also troubled by the promotion of a practice that is based on principles that cannot be supported by the established laws of chemistry, biology, or physics. Hopefully, the publicity the current lawsuit generates will help people understand the true nature of homeopathy.

Let's begin by explaining what homeopathy is not. It is not an umbrella term for alternative or complementary practices. The use of herbal medications or acupuncture or reflexology has nothing to do with homeopathy. Homeopathy is a specific practice conceived in the early nineteenth century by Samuel Hahnemann, a conventionally trained German physician, who became disillusioned with bloodletting, leeches, suction cups, purges, and arsenic powders, all standard treatments at the time.

It seemed to Hahnemann that these did more harm than good. He was probably right.

One remedy that did work was an extract of the bark of the South American cinchona tree used to treat malaria. But lacking standardized preparations, there was a problem with finding the right dose. Hahnemann, interested in the maximum amount his patients could tolerate, became his own guinea pig and began to take increasing doses of cinchona bark. He was surprised to find that, at a high dose, he developed symptoms much like the ones he witnessed in his malaria patients. At that epic moment, homeopathy was born! Hahnemann derived the term from the Greek "homoios," meaning "like," and "pathos," meaning suffering. "Like cures like," Hahnemann concluded. A substance that causes symptoms in a healthy person will cure like symptoms in a sick person when given at a smaller dose.

Hahnemann went further and began to systematically test the effects of a large variety of natural substances on healthy people. Such "provings" led him to conclude that belladonna, for example, could be used to treat sore throats, because it caused throat constriction in healthy subjects. But belladonna is a classic poison. Was homeopathy therefore dangerous? Not at all. Hahnemann had another idea. He theorized that his medications would work by the Law of Infinitesimals. The smaller the dose, the more effective the substance would be in stimulating the body's "vital force" to ward off the disease. This is a totally illogical conclusion.

"Active preparations" were made by repeated dilutions of the original extract. Hahnemann was not bothered by the fact that at these dilutions, none of the original substance remained; he claimed that the power of the curative solution did not come from the presence of an active ingredient, but from the fact that the original substance had in some mystical way empowered the solution with curative properties. A simple dilution, however, was not enough. The vial had to be struck against a special leather pillow a fixed number of times to be "dynamized" before adding a drop of the solution to a sugar pill.

These were bizarre ideas, to be sure, but Hahnemann must have been impressed by the success of his homeopathic treatments. No surprise here; the placebo effect can indeed be very impressive. And patients certainly preferred a treatment that had no side effects to being bled or being purged. A real curiosity was that Hahnemann did not advocate a homeopathic treatment for malaria using ultra diluted cinchona bark. He must have recognized this would not work.

Hahnemann didn't know about molecules, but today calculations readily show that homeopathic products such as Oscillococcinum 200C do not contain a single molecule of the duck organs that serve as the raw materials for the production of the final "remedy." The designation "C" represents an initial dilution of 1 to 100, and 200C means repeating this 200 times. "C" is confusing to the consumer because a larger number actually means a smaller dose, and in any case, the term does not conform to the Canadian Weights and Measures Act. This issue, while included in the lawsuit, is not its essence.

The main thrust of the legal action is that Oscillococcinum is mislabeled because the product clearly states that it contains the "medicinal ingredients" *Anas Barbariae Hepatis et Cordis extractum* 200C (duck liver and heart), as well as the non-medicinal ingredients sucrose and lactose. No chemical test can determine the presence of any "medicinal ingredient," and furthermore, the label states that every gram of product contains 0.85 grams of sucrose and 0.15 grams of lactose. For anyone, except perhaps homeopaths, 0.85 and 0.15 add up to 1, leaving no room for any other ingredient.

How can a product claim to contain a medicinal ingredient when no such substance can in any way be detected? Oscillococcinum amounts to a mislabeled sugar pill. If it is to be marketed, it should be honestly labeled. The lawsuit against Boiron and Shoppers Drug Mart aims to ensure that this happens, in order to prevent the public from being misled.

Homeopaths, of course, have to admit that there is not a single molecule of the original substance in the final product, but they maintain that the dilution and shaking leaves some sort of imprint on the solution.

When I dilute my chicken soup, its taste suffers. When I take one aspirin tablet instead of two, my headache doesn't resolve. When I use less detergent, my clothes do not come out as clean. Yet, in the topsy-turvy world of homeopathy, less is more. The more a biologically active substance is diluted, the more potent it becomes. The most powerful homeopathic drugs, the ones that have to be used the most carefully, according to some homeopaths, are the ones that do not even contain a single molecule of the original substance! Oscillococcinum, the purported cold and flu remedy made from the liver of a duck, falls into that category. At the declared homeopathic dose of 200C, the total mass of pills that would have to be consumed to encounter a single molecule of the original substance would be billions of times greater than the mass of Earth. Yet the label on this product says it contains a "medicinal ingredient"! And, curiously, it does not warn of the danger that such a "high potency" remedy presents.

Other homeopathic preparations may be derived from an astounding array of substances that include snake venom, fecal matter, arsenic, gold, plutonium, blister beetles, and the south pole of a magnet. Even more bizarre are "Light from Venus" and "Berlin Wall," a homeopathic dilution of which is supposed to help people with a lot of conflict in their lives.

Given that homeopaths have scientists breathing down their necks for an explanation of how nonexistent molecules can provide a therapeutic benefit, they have had to come up with some sort of a theory. The usual claim is that the process of dilution and "succussion" (banging the solution into a leather pillow between dilutions) "dynamizes" the remedy by leaving an "imprint" of the original substance. Chemists are prone to start pulling their hair when they hear something like that. Not to worry, though, homeopathy has a treatment for hair loss, *natrium muriaticum*. That's sea salt. But going for a swim in the ocean won't do; the salt is way too concentrated.

Can there be anything to the "water memory" idea? Water molecules do associate with each other momentarily through what any student of chemistry recognizes as hydrogen bonds. But these connections last only picoseconds before the molecules rearrange themselves. In any case, past a dilution of 30C, the solution contains no water molecules that have ever come into contact with the original substance! Furthermore, that original substance, as in the case of duck liver, is composed of thousands of different compounds. Which one is the water supposed to remember? And why does it not remember any of the other compounds it has encountered as it flowed through rivers and sewage systems? This, though, is hardly the point. Even if there were such a thing as water memory, why should that have anything to do with treating a disease? Homeopaths never address that question. They are too busy coming up with various pseudoscientific explanations about imprinting the memory of substances on water.

Another point that homeopaths seem to ignore is that their pills do not even contain any water! A drop of the diluted and "succussed" solution is added to a pill made of sucrose and lactose, but the water soon evaporates. So does it leave behind a ghost of the memory it supposedly contained? And how, exactly, is that ghostly memory released when the pill is swallowed and the sugar dissolves? Of course, if you are willing to abandon or misuse the laws of chemistry, physics, and biology, you don't have to concern yourself with such issues and can be satisfied by explanations that invoke "vital force" or "quantum entanglement."

Sometimes the effectiveness of homeopathy is likened to the effectiveness of vaccination. This is a hollow argument. First of all, vaccines contain measurable amounts of active ingredients. And we know how they work: they give rise to measurable amounts of antibodies. Furthermore, the active ingredients in vaccines are similar to the disease-causing agents. Homeopathic remedies contain no measurable ingredients, give rise to no measurable changes in the body, and the undiluted original "medicinal ingredient," such as duck liver, bears no resemblance in any way to the disease-causing organism, which in the case of a cold or the flu is a virus.

At a loss to explain how homeopathy works, homeopaths essentially invoke Hamlet's musings. "There are more things in heaven and earth . . . than are dreamt of in your philosophy." Seems appropriate, since both Hamlet and homeopaths appear to believe in ghostly images. Basically, the homeopathic argument comes down to, "we may not know how it works, but it works."

Homeopaths are convinced of the efficacy of their treatment because of the positive feedback they get from patients. But is this because their pills are effective, or is it because they tend to be caring people who listen to patients and spend a lot more time analyzing concerns than conventional physicians? Homeopaths will point out that there are proper randomized trials that show a benefit for homeopathy. Indeed, it would be shocking if there weren't any. When you carry out enough trials, some will, by chance alone, show a positive result. If you repeatedly toss 100 coins into the air, it won't be long before you come up with a result that differs significantly from fifty-fifty. That's why, instead of looking at individual studies, we rely on a meta-analysis, a study of studies.

Here, the results are clear. The effects of homeopathy are indistinguishable from the placebo effect. Not surprising, since homeopathic remedies are indistinguishable from each other. Or from sugar pills. The James Randi Educational Foundation offers a million dollars to anyone who can by any means identify an unlabeled homeopathic remedy. Certainly, any pharmaceutical company can readily identify any of their products. If this cannot be done for homeopathic remedies, how can a homeopath know he or she is giving the right substance? In fact, how can we differentiate between a real and a fake homeopathic remedy?

Critics of homeopathy have been known to swallow entire bottles of homeopathic pills to make the point they contain nothing but sugar. But homeopaths are not disturbed by this demonstration because, according to the tenets of homeopathy, increasing the dosage actually reduces the effect. So, the critics would face danger not by taking more pills, but by just licking one. Or, perhaps, they could overdose by staying away from the pills altogether.

We can safely say that homeopathic remedies pose no risk of side effects or of toxicity. Just try calling a poison control center to say that you accidentally took too many homeopathic pills. You'll get a response along the lines of "forget it" or "bogus product." But does this mean that homeopathy presents no risks? No, it doesn't. There are several concerns.

Some homeopathic remedies may not actually be homeopathic.

More seriously, some homeopaths offer pills for protection against malaria or radiation exposure. Others claim that they can treat cancer, with the most outrageous ones urging their victims to give up conventional treatment. Finally, there is the matter of Health Canada issuing a DIN-HM to homeopathic products, implying to the consumer that these remedies have been shown to be safe and effective. Safe, yes. Effective, no.

Let's amplify. Marketers sometimes use the term "homeopathic" to describe products that are not at all homeopathic. A classic case is Zicam, sold as an intranasal homeopathic cold remedy until 2009, when the Food and Drug Administration advised that the product be avoided because of a risk of damage to the sense of smell. How can a homeopathic remedy do that? Simple: it was mislabeled. Zicam actually contained a significant amount of zinc gluconate. This, though, is not nearly as serious as recommending ridiculous malaria protection pills that contain no active ingredient to people traveling to areas where the disease is endemic.

And how about "Homeopaths Without Borders"? I kid you not. Here is one of their gems: "With the onset of the rainy season in Haiti, there will be a great need for remedies to treat dengue, malaria, cholera, and other tropical diseases." Claiming that homeopathy can treat these diseases is criminal. Jeremy Sherr, of "Homeopathy for Health in Africa," goes even further: "I know, as all homeopaths do, that you can just about cure AIDS in many cases." Nonsense, of course, and even disparaging to most homeopaths, who draw the line at claiming cures for serious diseases.

Perhaps the most reprehensible practitioners of homeopathy are those who prey upon desperate cancer victims. The following comes from the "Wisconsin Institute of Nutrition," whatever that may be: "The important thing to know about cancer and choosing whether to use homeopathy or not is that surgery will not remove the disease. Most people will still opt for conventional treatment, so how can homeopathy be useful to them? They can take the appropriate remedy after surgery to prevent recurrence. For strict homeopathic thinking such a procedure is not optimum." Needless to say, there is zero evidence that sugar pills can prevent a recurrence of cancer.

Homeopaths are not ones to miss a marketing opportunity. Soon after the Fukushima nuclear power plant disaster in Japan in 2011, several offered remedies for either the treatment or prevention of radiation poisoning. Believe it or not, one of the suggested remedies was "X-ray." What is it? A sugar pill treated with a homeopathic dose of X-rays. I wonder how one dilutes X-rays. What bunk.

Scientists have always challenged homeopathy, but now consumers are beginning to realize the delusion of dilution. In California, homeopathic manufacturer Boiron settled a $12 million class-action lawsuit that alleged the company had violated false advertising laws by claiming that homeopathic remedies have active ingredients. As a result of the lawsuit, Boiron will now add a disclaimer stating that their products have not been evaluated by the U.S. Food and Drug Administration, and will also include an explanation of how the active ingredients have been diluted. In Australia, a woman is suing a homeopath whom she claims offered misleading information to convince her sister to give up conventional cancer treatment.

In Britain, the House of Commons Science and Technology Committee released a report stating that homeopathic remedies work no better than placebos and should no longer be paid for by the U.K. Health Service. The Committee also criticized homeopathic companies for failing to inform the public that their products are "sugar pills containing no active ingredients."

And at a British Medical Association Conference, an overwhelming vote supported a ban on any funding of homeopathic remedies, calling them "witchcraft."

In Canada, the Natural Health Products Directorate has a mandate "to ensure that Canadians have ready access to natural health products that are safe, effective, and of high quality." Yet it licenses homeopathic products without requiring proof of efficacy. Why should the manufacturers of these products be less accountable than those of other pharmaceuticals? Knowing this, how can pharmacists in good conscience sell sugar pills that claim to have ghostly images of molecules?

Homeopathic remedies work through the placebo effect. That, of course, is not negligible. Placebos can have success rates of over 30 percent! But if you think there's something more to homeopathy, consider the following: How come different homeopaths prescribe different remedies to the same person for the same condition? How come drugs other than homeopathic remedies do not increase in potency when they are diluted? How come trace impurities in the sugar used to make the tablets, or in the water or alcohol used for dilution, which are present at higher concentration than the supposed active ingredient, have no effect? How can remedies that are chemically indistinguishable from each other have different effects? And how come a producer of homeopathic remedies, if given an unidentified pill, cannot determine the original substance used to make the dilution? Finally, how come there are no homeopathic pills for diabetes, hypertension, or birth control?

I think I've said enough. According to the principles of homeopathy, if I say more and more about the irrationality of homeopathic remedies, the effectiveness of my arguments will become less and less.

Seeing Is Believing

Pictures don't lie, right? Of course they do. And they were deceiving us long before Photoshop made the manipulation of images almost child's play. Today, nobody would bat an eye at a ghostly image of Abraham Lincoln standing behind his grief-stricken widow, apparently comforting her. But back in the 1860s, when William Mumler produced the first "spirit photographs," the public was stunned. These photos appeared to show dead relatives hovering around the living subject who had posed for the picture. Photography was magical enough, so it didn't seem such a far stretch that the camera could see things that the human eye could not.

Mumler discovered "double exposure" accidentally when he mistakenly used a previously exposed but undeveloped photographic plate. He immediately recognized the financial potential of this discovery and reinvented himself as a psychic medium who specialized in communicating with the other side through photographs. By today's standards, his efforts were amateurish, but in the heyday of spiritualism, they were readily accepted as authentic. Only when Mumler made the mistake of using images of people who were still alive as his "ghosts," did his little scam crumble. But by this time, many other "spirit photographers" had recognized the lucrative nature of the business and had gotten into the game. And amazingly, the clever ruse even snared luminaries such as Sir Arthur Conan Doyle and Sir William Crookes. Conan Doyle, the creator of Sherlock Holmes, was a physician, and Crookes was a pioneer in chemistry and physics. One would think they would have known better.

Conan Doyle was a staunch believer in spiritualism, a position his famous detective would have taken a dim view of. But it was Sir Arthur's championing of another type of fake photograph that best demonstrates the extent of his credulity. In 1917, two young girls produced five photos that purported to show fairies dancing in the woods. Conan Doyle was convinced the pictures were real and refused to believe that he had been fooled by the simple trick of hanging cardboard cutouts by a thread in front of the camera. It was inconceivable to him that a couple of uneducated girls could pull one over on someone of his stature. The pictures therefore had to be evidence of the existence of fairies! In 1983, Elsie Wright and Frances Griffiths finally admitted that they had faked the photographs but nevertheless maintained that they had actually seen real fairies.

By the time the ladies had unburdened their souls, Roger Patterson and Robert Gimlin had outdone the "Cottingley fairies." In 1967, these two thrilled the world by capturing the first images of the fabled Bigfoot! Their short film shows a creature lumbering across the woods, looking very much like a man in a gorilla suit. There is good reason for that: it is a man dressed in a gorilla suit. The elaborate hoax was described in detail at a recent conference on magic history by Phillip Morris, a man who should know, since it was his costume company that provided and altered the gorilla suit used to stage the scene. Needless to say, there are legions of Bigfoot believers who don't buy Morris's claim and remain convinced that some sort of giant ape-like creature prowls the Pacific Northwest.

With such ample historical evidence about photographic manipulation, it's surprising how few people question the authenticity of a series of photographs being circulated on the internet purporting to show the results of a student's science fair experiment. The pictures depict plants supposedly watered either with microwaved water, or with water that has been heated on a stove top. And guess what! The microwave-watered plants wither while the others flourish!

One can come up with all sorts of possible explanations for the difference. Was the soil the same in the two plants? Were they given equal amounts of water? Could they have been exposed to different lighting conditions? Was there some difference in the seeds? But how about a simpler possibility: fraud. It isn't very hard to set up two plants side by side and ensure that one thrives while the other dies. Just water one and not the other. Of course, the possibility that this is the way the pictures were created does not prove the case.

Heating water in a microwave oven does nothing other than raise its temperature. Any talk about "the structure or energy of the water being compromised" is plain bunk. But absurdly implausible arguments don't prove that the pictures are fraudulent, either. What proves it is the good old standard of science? Reproducibility! Or lack thereof.

I did the experiment. I watered plants with microwaved water, kettle-boiled water, and stove-top boiled water. I felt pretty silly about it, but I did it. The results? As expected, no difference. I didn't take any pictures because, after all, how would you know that they are not faked? So here is the choice: you can take my word that the experiment cannot be reproduced, accept that science tells us that microwaves do nothing to water other than heat it, or take at face value some pictures in a circulating email that purport to show an effect that has eluded scientists around the world but was discovered by a student pursuing a science fair project. Better yet, do the experiment yourself!

As you might guess, I don't believe in spirit photographs, fairies, Bigfoot, or plants succumbing to the vileness of microwaved water. And I would have put goats that climb trees into the same "unbelievable" category. But I would have been wrong. It seems that some Moroccan goats have learned to climb the argan tree in search of its olive-like fruit. Legend has it that the undigested seeds that pass through the goats used to be collected and pressed into "argan oil," a traditional food flavoring. Highly questionable. The oil, also used in the cosmetics industry, is actually pressed from fruit that has been picked by human hands, making the tree-climbing goats a nuisance. Still, one can appreciate their remarkable athleticism. It's easy to find pictures of their exploits online. And pictures don't lie, right?

A Look at Braco the Gazer

What a clever scheme! There's no overt deception. That's because you don't claim to be able to do anything. You don't preach. You don't offer any sort of philosophy. In fact, you don't even talk. You don't touch anyone. You don't sell any potions. You don't use any sleight-of-hand tricks. You don't use any sort of equipment. You don't wear strange clothes. However, you do grow your hair to project an image of a certain biblical figure associated with healing. But you don't call yourself a healer, although you do not object if others do. In fact, you do nothing but promise to gaze at people for about seven minutes if they plunk down eight dollars. You are "Braco the Gazer." And you are a phenomenon!

Picture this: thousands of people flood into an auditorium, many looking ill, some hobbling with canes, others in wheelchairs, reminiscent of crowds that flock to faith healers, ready to open up their pocketbooks in return for a few miracles. But in this case, there are no promises of miracles. Not directly, anyway. As the crowd buzzes with anticipation, the proceedings begin with the session's host welcoming everyone to the meeting with "the healer who doesn't call himself a healer." A nice little legalistic "out." Everyone's experience will be different, the audience is told, and "skeptics will become believers." "There should be no specific expectations." But of course there are. People have heard that Braco's silent holistic gift can clarify the mind, make pain vanish, and wither tumors. It can also repair stalled cars and stop cats from vomiting.

There are a few instructions before the holy man, who does not claim to be one, appears. Cell phones and other electronic devices must be turned off because they may disrupt Braco's "energy," despite the fact that he himself makes no claim to projecting any such thing. Then, a warning: the session is only for people over the age of eighteen, because for youngsters the gazing energy is too powerful. Ditto for women who are more than one trimester along in their pregnancy. That's a curious one, because developmental problems are most likely to be initiated in the first trimester. An exception is made on November 23, Braco's birthday, when families can bring children. Perhaps on that day he tones down the energy that he makes no claim to have.

The host's introductory remarks are followed by a video of an unfortunate skeptic who had been diagnosed with "Agent Orange cancer virus" (a ridiculous and befuddling term) and had attended a previous event with the healer who does not claim to be a healer. The skeptic went home, his idea that this was all bunk confirmed. But two days later, a blood test declared him to be cured! (Must be some blood test, capable of detecting a nonexistent virus.) After a few more words about the importance of being skeptical, and instructions to hold up photos or X-rays of sick people to be cured in absentia by the man who claims no healing ability, the time arrives for the "Silent Gaze."

Braco, the Croatian non-healing healer, has been enthralling massive audiences in Europe for some eighteen years, but only in 2010 did he discover the greenback pastures of America. In

Europe, he usually limits himself to just one gazing session per day, but everything is bigger in America. Here visitors can cycle through the lines of "Braco Gazing" all day long, as long as they pay an entry fee each time. And for this all they get to do is gaze at the gazer. Braco struts onto the stage, long hair flowing, face expressionless. You wouldn't be surprised to hear "Jesus Christ Superstar" bursting from the loudspeakers, but all you hear is some New Age music. Body almost motionless, he . . . well . . . gazes. That's what gazers do. They gaze. And gaze. Some of the "gazees" snicker, others revel in rapture, curiously with their eyes closed. Maybe the magical gaze penetrates eyelids.

After about seven minutes, it's over. He glides off the stage, the room empties, ready to be refilled by a new throng, along with some repeaters who feel they need another dose of healing energy from the man who makes no claim to have any. In the lobby, there are testimonials galore about toothaches disappearing, back problems vanishing, and bodies being filled with intense heat. But those who came in wheelchairs leave in them. One lady claims to have been overcome by a "big bubble of love." It is not exactly clear what this means, but she seems to have been "satisfied." There are books and DVDs to buy, as well as jewelry that features a thirteen-pronged star. Again, no claims are made other than that the Sun is the symbol of life and the Sun is the source that gives us life, light, and energy. Can't argue with that.

Of course, not everyone can get to one of Braco's events. That's no problem, though, because thanks to modern technology, you can experience the gaze through live streaming. At $3.00 per session, it seems a bargain. That's a lot less than what one might spend on various dietary supplements, magnets, crystals, pendulums, power bracelets, aerobic oxygen solutions, or homeopathic preparations that are marketed with testimonials virtually identical to those heard from people who have been gazed at by Braco.

I thought I'd give the silent gazer a look. He is a good gazer; I'll give him that. But there was no heat, no infusion of vitality, no sensations of inner peace, no awakenings of consciousness, just some thoughts about what he was thinking about as his gaze delivered its dose of placebo. Maybe I should have stuck with it for the full twenty minutes I paid for. Maybe it's a dose-dependent thing. I can't complain, though. Unlike detox foot pads, Kangen water, or zero point energy healing wands that do not live up to their lofty promises, Braco gives you exactly what you paid for. He will gaze at you for the period of time you purchased. He's a clever man. A lot more clever than the folks he gazes at.

Celebrities and Cerebral Claptrap

Suzanne Somers is a former actress with a pretty face and, probably, firm thighs. After all, she did advertise ThighMaster for years. As far as her smooth skin goes, she offers an electrifying explanation. She uses a gizmo that sends tiny jolts of electricity to give her facial muscles a "workout." Right. Unfortunately, Suzanne has had to deal with problems that go beyond her thighs and wrinkles. She has had a bout with breast cancer. But neither her good looks nor her struggles with illness qualify her for donning the mantle of a scientific guru. Yet, that is just what she has become. And guruhood means that Suzanne influences many lives as she sounds off on diets, "bioidentical hormones," nutritional supplements, and "alternative" cancer treatments. Her success in beating breast cancer, she claims, is linked to injections of mistletoe extract. Never mind that she had a lumpectomy and radiation treatment.

In Suzanne's book *Knockout*, physicians like Dr. Nicholas Gonzalez are placed on a pedestal because they are "curing cancer" through unconventional means. Except that the "cures" are not supported by facts. Gonzalez's regimen of numerous dietary supplements and coffee enemas has actually been tested by the National Center for Alternative and Complementary Medicine, an organization not exactly adverse to alternative therapies. Patients fared more poorly than those on conventional chemotherapy. But for Suzanne, this doctor, who has been reprimanded by the New York state medical board for "departing from accepted practice," who was forced to submit to psychological examinations and undergo retraining, and who has lost malpractice suits in which he was accused of negligence in cancer treatment, is a hero to whom we should listen.

In *Sexy Forever*, Suzanne expands her scientific expertise to gene expression. In an interview about this book, she informs us that, "We have five cancer protective genetic switches in our bodies that get turned off by diet and lifestyle. One is turned off by toxins and chemicals; one by poor quality food, i.e., non-organic, pesticide-laden food; one by lack of sleep; one by stress; and one by imbalanced hormones. Now, what women really need to understand is that, first and foremost, to turn back on your protective genetic switches, you've got to get the hormone switch turned back on. Imbalanced hormones are a big factor in why women are fat, and when women get fat they get very unhappy."

How does such pseudoscientific blather get Suzanne on talk shows? We are talking about a lady who was insulted when her oncologist asked why she was taking steroids. "Who, me?

Never!" Suzanne, it seems, had no idea that the bioidentical hormones she was taking were steroids. And let's not even mention the folly of self-treatment of an estrogen-receptor-positive form of breast cancer with estrogenic hormones, which is exactly what Suzanne's bioidentical hormones are.

Why then do people regard her as an authority on beauty, weight loss, and health? Because she is a celebrity! And we have a culture of celebrity worship. Especially if they look good, to boot. Talk-show sets have become lecture rooms, books written by celebrities are the new texts, and many learn their science from Professors Somers, Tom Cruise, Demi Moore, Julia Sawalha, Sarah Palin, Michelle Bachmann, and Alex Reid.

You've probably not heard of Alex Reid, but this martial arts fighter and actor is a big name in Britain. He makes the rounds of talk shows, dates glamorous models, has been in movies and soap operas, and has even had his own TV show. How does he prepare for fights? Let's listen to his philosophy. "It's actually very good for a man to have unprotected sex as long as he doesn't ejaculate because I believe that all the semen has a lot of nutrition. A tablespoon of semen has your equivalent of steak, eggs, lemons, and oranges. I am reabsorbing it into my body and it makes me go raaaaahh!" Makes me go nuts. Give us a break, Alex!

Let's not even address the physiological and nutritional nonsense, but the bit about unprotected sex is dangerous claptrap. Talking about dangerous claptrap, how about British actress Julia Sawalha's views on traveling to places where malaria is endemic. "I don't get inoculations or take anti-malaria tablets, I take the homeopathic alternative called 'nosodes,' and I'm the only one who never comes down with anything." Lucky Julia. Perhaps she can point us toward some trials that demonstrate how preparations with no measurable active ingredients, which is the case for homeopathic "drugs," can prevent malaria.

Demi Moore also has views on health. She thinks it can be optimized by treatment with leeches. "They have a little enzyme and when they are biting down on you, it gets released in your blood and generally you bleed for quite a bit and then your health is optimized — it detoxifies your blood." Ridiculous. So is Tom Cruise's declaration that "psychiatry doesn't work; when you study the effects it's a crime against humanity." According to Scientology, which Cruise espouses, our problems are caused by mental implants we received from space aliens and should be treated by Scientology's mysterious "auditing" methods. So who is committing the crime against humanity?

Professor Sarah Palin once lectured us on research funding. "Sometimes research dollars go to projects that have little or nothing to do with the public good. Things like fruit-fly research. I kid you not." This is breathtaking ignorance. Fruit flies are excellent models for the links between genes and disease and have provided important clues for exploring numerous conditions ranging from birth defects to autism. And then there is former presidential hopeful Michelle Bachmann regaling us with her toxicological knowledge. "Carbon dioxide is portrayed as harmful," she says, "but there isn't even one study that shows carbon dioxide is a harmful gas." No, Professor Bachmann, it isn't harmful; we inhale and exhale carbon dioxide all the time. But the issue is the role that carbon dioxide plays in global warming.

Let's conclude with model Heather Mills, who offers insight into our own demise. "Meat sits in your colon for forty years and putrefies and eventually gives you the illness you die of. And that is a fact." No, Heather, it isn't.

Rhinoceros Horn *Is* Useful — for Its Original Owner

Ignorance can be lethal. Especially if you happen to be a rhinoceros. You may end up being murdered on account of the senseless notion promoted by some Traditional Chinese Medicine (TCM) practitioners that the horn can treat a host of ailments ranging from arthritis, delirium, and fever to food poisoning and devil possession. This foolishness has been lurking around for over 2,000 years but, needless to say, long-term use does not prove efficacy. There is no evidence that rhinoceros horn is of any use to anyone except its original owner.

Drinking a brew made from powdered horn has never cured anyone of anything. That can be said with some confidence because pharmaceutical companies, as well as scientists at the Chinese University of Hong Kong, have actually looked into the horn's biological properties. The only effect noted was a lowering of the rectal temperature in rats that had previously been treated with turpentine to induce fever. Even that "remedy" required a dose some 100 times greater than that recommended for humans, according to the tenets of TCM. The lack of efficacy is not surprising given that the horn is made of keratin, the same protein found in animal hooves and fingernails. Indeed, ingesting rhino horn has about the same therapeutic effect as chewing on fingernails. But it is considerably more expensive!

A hundred grams of horn can sell on the black market for thousands of dollars. And prices have been increasing ever since "cancer cure" was added to the rhino horn's wondrous properties after a story began to circulate about a Vietnamese politician's supposed miraculous recovery from the disease. Vietnam has now become a key market for rhino horn smuggled out of Africa, with most of the horns coming from poaching. But the South African government also issues a limited number of licenses for "trophy hunting." It seems, however, that some of these licenses have been issued to Vietnamese "hunters" who do not even know how to hold a rifle. The rhinos are actually being shot by their African guides, a practice that is expressly forbidden. And the horns, instead of ending up on mantles, are sold illegally on the street, for exorbitant amounts, as medicine.

The South African government tries to prevent such sales by putting a microchip into horns that have been obtained through legal hunting, hoping to follow their path. This has proven to be of questionable efficacy. Rangers also engage the poachers, with gun battles being a common occurrence. There are, however, also reports of collaboration with the criminals in return for tidy payments.

In addition to the remarkable therapeutic properties of rhino horn, stories have also long circulated in the West about its effectiveness as an aphrodisiac. These took wings soon after Viagra debuted in 1998, precipitating an onslaught of media coverage. One angle that made the rounds was based on research published by Dr. Bill von Hippel, a psychologist from the University of New South Wales, and his brother, Dr. Frank von Hippel, a biologist at the University of Alaska. The rhinoceros, reporters implied in reference to the Hippels' research, should be thankful for Viagra. Why? Because Asian men were switching to the drug from rhinoceros horn, a "traditional aphrodisiac" for which rhinos were being hunted to extinction. But there are several problems with the story.

First, the von Hippels never suggested a connection to the rhinoceros. Second, Traditional Chinese Medicine has never regarded rhino horn as an aphrodisiac. The von Hippels investigated the decline in sales of seal penises and reindeer antlers, both of which are traditional treatments for impotence.

According to their research, sales plummeted after the introduction of Viagra, and surveys in Hong Kong apothecary shops confirmed that traditional Chinese medicines for impotence were taking a modest hit. That was interesting because there had been no previous evidence of Chinese customers switching to Western medicines for other common ailments such as headaches or indigestion. But the von Hippels suggested that failure to achieve an erection isn't comparable to a headache. Unlike with seal penises or deer antlers, the effects of Viagra are rapid and visible to the naked eye. The overall conclusion was that Viagra might account for a reduced demand for several animal species that are over-harvested for the purpose of treating impotence with Traditional Chinese Medicine. It was the press, not the scientists, who factored the rhinoceros into this equation.

There's yet another glitch to the claim that Chinese men are giving up rhino horn in favor of Viagra. Besides the fact that there is no tradition of using it as an aphrodisiac, the sale of rhino horn for medicinal purposes is illegal in China. At least, for now. That may change because proponents of TCM are pushing for legalization, maintaining that rhino horn can cure everything from brain hemorrhage to AIDS. A prominent advocate claims that the horn isn't being used because government officials have been "tainted by Western thought."

Actually, they may be tainted by another form of Western thought: the promise of profits. It seems China has been quietly importing rhinos supposedly to stock "Africa View" tourist parks. The scuttlebutt is that the parks are actually a front for the Longhui pharmaceutical company that wants to breed the animals and turn their horn into medicine, apparently without killing the animals. The company has developed "live rhino-horn grinding technology" that allows horn to be harvested from live animals. Like fingernails, the horn regrows. Longhui hopes that this technology will allow for the legal sale of rhino horn and argues that it will also keep rhinos from being mercilessly killed by poachers.

This seemingly shiny effort is actually tarnished. Legalizing the sale of the horn would give implicit approval to its value as medicine. It has none. Promoting nonsense is never the right approach. The way to protect the rhino is by spreading the word that reliance on ancient authority is not the path to take. Evidence-based science is! While you're at it, why not protect other endangered creatures too by explaining that tiger bone, deer antler, seahorse powder, and bear bile can all be replaced by medicines that actually work. And if you find that your rats' rectal temperature is elevated, reach for aspirin, not rhino horn.

The Skinny on the HCG Weight Loss Scheme

"Why be stout," queried a newspaper ad in 1927, "when all you have to do to bring back natural slimness is rub your body with Dr. Bouchard's Flesh Reducing Soap? It will absorb all fatty tissues from any part of the body and take away from large hips, double chin, ungainly ankles, arms, legs, bust, and waistline." Readers were assured that it was "perfectly safe and proved that dangerous drugs, dieting, steam packs, or exercises were needless." Also in 1927, researchers discovered a hormone in the urine of pregnant women that stimulates the production of progesterone, which in turn prompts the production of blood vessels in the uterus needed to sustain a growing fetus. That hormone, human chorionic gonadotropin (HCG), would form the basis of the first pregnancy test. It would also become useful as an ovulation inducer for women experiencing fertility problems. However, it was as a purported weight control miracle with claims rivaling Bouchard's Soap that HCG first flashed into the public eye.

P.T. Barnum famously claimed that there is a sucker born every minute. But Kevin Trudeau, the king of infomercial scams, is making a valiant attempt to prove that Barnum was an optimist. Overweight people are an inviting target for the crafty Trudeau, who has managed to resurrect a mercifully forgotten diet plan that had been introduced in the 1950s by British endocrinologist Albert T.W. Simeons. Injections of HCG, coupled with a 500-calorie-a-day diet, lead to effective weight loss, claimed Simeons.

Dr. Simeons had been investigating human chorionic gonadotropin as a treatment for Frohlich's syndrome, a delayed development of the genitals in adolescent boys caused by a pituitary gland disorder. The term "gonadotropin" actually means "stimulating the gonads," and some athletes have even been known to use it to counter testicular shrinkage due to steroid use. Baseball star Manny Ramirez was suspended for fifty games in 2009 due to his use of the drug. Although there was no proof that Ramirez had been using steroids, the presence of HCG in his urine was a red flag. It is common for steroid users to use HCG both during and after a steroid cycle to boost testosterone production and restore testicular size.

It was precisely HCG's effect on the genitals that had intrigued Simeons, but during his trials he also noted that boys treated with the hormone had a reduced appetite and lost weight. Surprisingly, they never complained of hunger! That was enough for Dr. Simeons to refocus his research on the possible use of HCG as a drug for weight loss. There was some theoretical justification for this since during pregnancy HCG is known to mobilize fat from areas in the body where it is stored, such as the hips, belly, thighs, and derrière, in order to ensure that the developing fetus is well nourished even if mom doesn't eat right.

Of course, any weight-loss regimen must limit caloric intake. But Simeons contended that with elevated levels of HCG in the blood, even severe calorie restriction would not cause hunger. And his restriction was severe indeed — 500 calories a day! That's basically starvation. Simeons argued that the extra calories necessary to fuel the body would be derived through metabolization of the fat that was being released from fat stores. It seemed as though the Holy Grail of weight loss had been found!

Other researchers weren't so sure. Nor was the U.S. Food and Drug Administration, which declared that all advertisements for HCG injections must include the disclaimer that "this drug was not approved as a safe and effective treatment for weight loss." Still, the fact was that people who received daily injections of HCG and maintained the 500-calorie diet were losing significant weight. But was it the hormone that was keeping the hunger pangs away? Scientists were sufficiently intrigued to mount a number of studies. By 1995, fourteen randomized double-blind controlled trials had been published. The conclusion of a meta-analysis of all the trials was emphatic: "There is no evidence that HCG is effective in the treatment of obesity; it does not bring about weight-loss or fat-redistribution, nor does it reduce hunger." Basically, the researchers declared, HCG is an effective placebo.

As a result, HCG disappeared from the radar until Trudeau resurrected it in his "They Don't Want You to Know" series. "They" apparently constitute some sort of nebulous alliance between industry, the medical establishment, and government, who for their own nefarious reasons want to keep people sick and fat. One would think that endorsement by a felon with multiple convictions and millions paid in fines would not amount to a marketing success. Never mind that the Federal Trade Commission even charged Trudeau with misrepresenting the contents of his book in his infomercials, in which he claimed that the HCG weight-loss plan is easy and safe. Despite all of this, Trudeau's books have been incredibly successful.

Trudeau's popularization of HCG resulted in the sprouting up of clinics offering injections and hucksters selling "homeopathic" versions of the miraculous weight-loss product. Homeopathic HCG drops are a total scam because they contain no HCG at all. In any case, HCG is a protein and would be quickly broken down in the digestive tract if taken orally. Both the FDA and the Federal Trade Commission (FTC) have issued letters to several companies warning them that they are selling illegal homeopathic HCG weight-loss drugs that make unsupported claims. Canada should follow suit.

Since HCG is a legal prescription drug, albeit for infertility, physicians are free to prescribe it as they choose. And for those who choose to prescribe it in this fashion, there's a tidy little profit. The safety of HCG in terms of regular injections for weight loss has never been established, and it is curious that people who are worried about trace amounts of endocrine-disrupting chemicals in the environment are willing to pump a known hormone into their body on a regular basis.

While the risks of HCG may still be unknown, the risks associated with a 500-calorie diet are well established. Gallstones, irregular heartbeat, electrolyte imbalance, nausea, hair loss, and fatigue are just some of the delights of a starvation diet that consists of coffee and an orange for breakfast; a little fish and raw asparagus for lunch; a piece of fruit in the afternoon; and a dinner of crab, spinach, Melba toast, and tea. Little wonder there are legions of dieters who boast about quickly losing 20 to 30 pounds. That's what starvation diets can do. But where are the people who have taken this road and have managed to keep the weight off after a year?

It all comes down to a basic question: when it comes to dieting and health, should we trust the peer-reviewed literature, the FDA, Health Canada, or should we listen to an infomercial mogul with a reputation dirty enough to warrant a search for Dr. Bouchard's Flesh Reducing Soap? Maybe I shouldn't even mention it, though, because Kevin Trudeau might try to resurrect that folly as well.

Doctors Who Kill

Some practitioners today still recommend lobelia as a "blood cleanser" and as a respiratory stimulant to treat asthma, while Boiron laboratories market Lobelia inflata as a homeopathic remedy to help people wean themselves off smoking. Unlike with the herbal preparation, there is no concern here about side effects, since the homeopathic remedy has been diluted to an extent that it contains essentially no lobelia. As we have already seen (page 33), while Thomsonism as such has been relegated to the history books, the pukeweed doctor's legacy of eschewing schooling and science in favor of reliance on "intuitive wisdom" and "nature's pharmacy" is unfortunately still with us.

> I will prescribe regimen for the good of my patients according to my ability and my judgment and never do harm to anyone. To please no one will I prescribe a deadly drug, nor give advice which may cause his death . . .

That passage comes from the Hippocratic Oath, by which physicians promise to practice medicine ethically and honestly. Not all abide by the oath, but doctors who willfully harm their patients are rare. The ultimate harm, of course, is murder. It is a crime for which Dr. Harold Shipman paid with his life, and for which Dr. Michael Swango will spend the rest of his days in jail. And it's one with which Dr. John Bodkin Adams apparently got away.

Adams practiced general medicine off and on from the 1920s to the 1980s in Eastbourne, England. At one time he was reputed to be the wealthiest general practitioner in the country, but gossip around Eastbourne had it that Adams was actually an "angel of death" who relieved wealthy widows of suffering as well as of their bank accounts. The word was that the doctor was a kindly practitioner, at least until he was able to persuade a wealthy widow to include him in her will. Once that was secured, the lady soon drifted off into the other world in a morphine- or heroin-induced stupor.

The large number of wills in which former patients named Adams as a beneficiary eventually prompted a police investigation that led to an arrest and trial. While he admitted to using narcotics to ease patients' suffering, Adams vehemently denied deliberately overdosing them. His inclusion in their wills, he claimed, was in appreciation of the excellent care he provided. The police had a different view. Of the 310 death certificates signed by Adams, 163 were deemed to be suspicious. But much of the prosecution's case was based on testimonials from nurses who claimed to have witnessed the doctor administering drugs in lethal doses. The defense, however, managed to produce notebooks in which the drugs administered had been carefully recorded, sometimes by the very nurses giving evidence in court. The notes did not match the nurses' recollection of what happened, and the jury decided that guilt had not been proven. Adams was later convicted of forging prescriptions, lying on cremation forms, and failing to keep a dangerous drugs register. He was fined £2,200, which he had no problem paying since he was receiving legacies from patients' wills up to his death at age eighty-four in 1983, with suspicions of having killed 160 people still hanging over his head.

When it comes to Dr. Harold Shipman, there's no question of suspicion. He murdered patients, possibly as many as 260. Shipman began practicing medicine in West Yorkshire in 1974, and before long, he was practicing murder as well. But it wasn't until 1998 that the police were alerted to the high death rate among his patients and the large number of cremation forms he had signed. Shipman was finally arrested after forging the will of an elderly patient whose body was subsequently exhumed and found to contain heroin. The doctor was eventually found guilty of fifteen murders and was sentenced to life with no possibility of parole. In 2004, he committed suicide by hanging himself with bed sheets from the window bars of his cell.

Mystery still surrounds Dr. Shipman's long murderous spree. Was he mad? Was he obsessed with the power of dispensing death? Did he want to relieve suffering? Was it for financial gain? We will never know. But we do know why Dr. Michael Swango became a serial killer. He simply enjoyed holding peoples' lives in his hands and terminating them at a whim.

Swango graduated from Southern Illinois University School of Medicine in 1983, and used his privilege as a physician to poison as many as sixty people, killing about thirty-five. His undergraduate career already smacked of an interest in poisoning, exemplified by a chemistry paper he wrote dealing with the 1978 case of Georgi Markov, a Bulgarian writer who was assassinated in London by the Bulgarian secret police.

Markov was poked in the thigh with a spring-loaded umbrella that injected a tiny capsule containing the natural toxin ricin, extracted from castor beans. Possibly it was this unusual murder that whetted Swango's appetite for eliminating people in an ingenious fashion.

Dr. Swango's trail of death began during his residency, when an unusual number of patients in his care met untimely ends. His colleagues nicknamed him Double-O Swango, a reference of course to James Bond, whose double O number gave him a license to kill. Suspicions were that Swango was injecting patients with a potassium chloride solution, leading to a quick death that was hard to detect. Nothing was proven, but Swango's hospital privileges were terminated. He then found employment as an emergency medical technician.

It wasn't long before some of his coworkers got sick with stomach cramps after Swango had treated them to donuts. They got suspicious and looked in the doctor's bag, where they found a bottle of Terro ant poison, which in those days contained arsenic. That's when they decided to spring a trap by leaving a pot of iced tea where they were sure Swango would notice it. The next day, they sent the tea to an FBI lab where, sure enough, arsenic was detected.

A police search of Swango's apartment revealed all sort of potential poisons as well as a book called *The Poor Man's James Bond* about do-it-yourself murder. Swango was sentenced to five years in prison but was out in two and a half. Despite having lost his medical license, he was able to find positions as a doctor in hospitals in the U.S., Saudi Arabia, and Africa. Using epinephrine or succinylcholine chloride, both very hard to detect, Swango killed a number of other patients until he was finally locked up without the possibility of parole in 2000. Part of the evidence at his trial included a passage he had copied from a book about a serial killer physician called *The Torture Doctor*: "He could look at himself in the mirror and tell himself that he was one of the most powerful and dangerous men in the world . . . he could feel that he was a god in disguise." Swango was no god; he was more like the devil incarnate.

GRAY

Fishy Claims for Fish Oil Supplements

It's a pretty common scenario: an observational study suggests that some food or beverage is associated with some aspect of health. A hypothesis is forged about the effect being due to some particular component. The component is isolated and tested in cell cultures or in animals with some intriguing results. A few small-scale human studies follow and generate optimism. Supplement manufacturers gear up and begin to flood the market with pills containing the supposed active ingredient. Their ads are supported by references to cherry-picked data, their hype couched with many a "may." Personal testimonials of benefit pour in and profits mount. The results of randomized, controlled double-blind trials (RCTs) begin to emerge, with contradictory results. Marketers highlight the positive results and dismiss contradictory research as "flawed." Sales continue to increase with various producers claiming that their product is superior to that of competitors.

As controversy mounts, researchers undertake meta-analyses that pool data from the best available studies. Results suggest that the initial optimism cannot be supported but there is a call for more studies. (Isn't there always?) Sales begin to slump as manufacturers and their industry associations scramble to punch holes in the meta-analysis and issue press releases that emphasize the studies with positive findings. Regulatory agencies walk a fine line, having to take into account business interests, freedom of choice arguments, and public health. Consumers are left bewildered, not knowing whom to believe or trust. In recent years we have seen such scenarios unfold with vitamin E, calcium, beta-carotene, ginkgo biloba, and now, omega-3 fats.

The omega-3 saga can be traced back to the early 1970s, when Danish researchers discovered a surprisingly low incidence of heart disease in Inuit tribes despite a diet dominated by fatty fish. Could this be due to the specific type of fat found in fish, they wondered? After all, the molecular structure of these fats differed from the fats found in meat and most vegetables. Perhaps these polyunsaturated fats, which feature a double bond on the third carbon from the end of the molecule, the so-called "omega carbon," had some special cardioprotective property. This was a reasonable guess, given that other populations around the world that consumed a fish-rich diet, such as the Japanese, were also known to have a low incidence of heart disease.

Laboratory studies soon revealed that the omega-3 fats have anti-inflammatory properties and can also reduce the clotting tendency of blood. Both of these observations mesh with theories of reduced cardiac risk. There were also implications of reduced blood pressure and a slowing of the progression of arteriosclerosis. Then came evidence of a lowered incidence of abnormal heart rhythms in fish consumers and a decrease in triglycerides in their blood, both established risk factors for heart disease.

By the 1980s, supplement manufacturers had begun to capitalize on the tantalizing studies and were filling capsules with various mixtures of docosahexaenoic acid (DHA) and eicosapentaenoic acid (EPA), the two dominant fatty acids in fish oils. Recommended dosages were no more than educated guesses. The U.S. Food and Drug Administration essentially endorsed the supplements, allowing labels to state that "supportive but not conclusive research shows that consumption of EPA and DHA may reduce the risk of coronary heart disease." Furthermore, the FDA approved a high-dose prescription mixture of EPA/DHA derived from fish oil for the treatment of high levels of triglycerides.

Before long, claims of protection against heart disease were joined by a plethora of others. Omega-3 fats were said to reduce the risk of cancers of the colon, breast, and prostate as well as that of macular degeneration and gum disease. They were also said to be useful in the treatment of depression and anxiety. There were claims of a slowing of cognitive decline in the elderly and improvement in attention deficit hyperactivity disorder (ADHD) in children. Moms consuming fish oils supposedly gave birth to children with higher IQs. Even pets benefited from fish oil supplements, sporting shinier coats. As the twentieth century came to an end, we were swimming in a sea of claims about the wonders of omega-3 fats. We were hooked on fish oil.

And then the double-blind studies started to appear and the scales began to fall from our eyes. Suddenly we were confronted with headlines such as: "Fish Oil Disappoints Versus Cancer," "Fish Oil Won't Fix Abnormal Heart Rhythms," and "Omega-3s of No Added Benefit to Heart Attack Patients." The highly respected journal *Circulation* featured a study demonstrating that among heart attack survivors, 1,000 milligrams of purified omega-3 oils per day for one year was no better than olive oil at preventing sudden cardiac arrest, death, heart attack, stroke, or the need for bypass surgery or angioplasty. The *British Medical Journal* (BMJ) published a study that found survivors of a heart attack or ischemic (clot-caused) stroke, or those with unstable angina (chest pain at rest), taking 600 milligrams of omega-3s a day for almost five years to be no better off than with a placebo at reducing nonfatal heart attacks, strokes, or deaths from cardiovascular disease.

Still, business was going along swimmingly for the fish oil supplement industry until the publication of a recent meta-analysis in the *Journal of the American Medical Association* that pooled the results of twenty high-quality randomized trials involving 68,000 people and found that supplementation with omega-3 fats did not reduce the risks of all-cause mortality, cardiac death, sudden death, heart attack, or stroke. The researchers conclude that their findings "do not justify the use of omega-3 as a structured intervention in everyday clinical practice or guidelines supporting dietary omega-3 fat administration."

Needless to say, the industry responded, claiming the analysis was flawed since many of the studies were on people who were already ill and therefore might not apply to people who were taking supplements just to maintain health. Furthermore, most studies, they claim, didn't control for the amount of fish people were already eating. If they already were consuming a fair amount, the supplements would not be expected to have an effect. Given the size and number of the trials embodied in the meta-analysis, these seem to be weak counterarguments. The analysis did, however, provide some comfort for people taking omega-3 supplements in that no harmful effects were noted. It turns out, as it almost always does, that eating the whole food is better than gulping the fishy claims made on behalf of some supplement. But unfortunately people are tantalized by simple solutions to complex problems and are far too ready to swallow such claims hook, line, and sinker.

Swallowing Blueberries, Apples, and Hype

Blueberries may reduce the growth of breast cancer! Apples and pears reduce the chance of stroke! I bet I have your attention now. But those are not my words; they're recent newspaper headlines. It seems that virtually every day some new study comes out touting the ability of this or that food to extend our earthly existence. Usually, the researchers themselves are modest in their claims and end their discussion with the inevitable call for more research. But then the media get a sniff of the action. And in the drive to capture public attention, science sometimes takes a backseat. Before long, a smidgen of science may be blended with a dash of hope and a healthy dose of hype to cook up a scrumptious headline. But for the scientifically minded, the tasty headline may trigger a bout of mental indigestion.

The blueberry story is a report of an interesting study carried out on female nude mice. Don't get any mental images of Minnie enticing Mickey; these nude mice are specially bred for laboratory research. They derive from a strain with a genetic mutation that causes them to have an underactive thymus gland, resulting in an impaired immune system. Outwardly, they lack body hair, hence the nickname "nude." Suffice it to say that these nude mice are not a perfect model for predicting biological effects in other mice, let alone in humans. Still, they are valuable in research because cancer cells can be introduced without a rejection response. And the blueberry study was all about injecting mice with breast cancer cells. But these were very specific breast cancer cells, known as triple negative cells.

The "triple negative" refers to the fact the growth of these cells is not supported by the hormones estrogen or progesterone and that they also test negative for the presence of human epidermal growth factor receptor 2 (HER2), a protein that promotes the growth of cancer cells. Triple negative cells therefore do not respond to standard hormone-blocking drugs such as tamoxifen or to medications such as herceptin, which interfere with the HER2 receptor. Triple negative cells are very aggressive, but are the cause of only 15 percent of all breast cancers.

Now, back to our blueberries. Researchers at City of Hope Hospital in Los Angeles treated nude mice to a diet that included either 5 percent or 10 percent blueberry powder by weight. After two weeks, the mice were injected with the triple negative breast cancer cells. A control group of animals was fed in the same fashion but without the blueberry powder.

Why undertake such an experiment? Because earlier laboratory studies had shown that blueberry extract had anti-angiogenesis activity, meaning that it interfered with the formation of blood vessels that tumors need to grow. After six weeks, the mice fed the 5 percent blueberry diet had a tumor volume that was 75 percent lower than the control animals, but strangely, those fed the higher dose blueberry diet showed only a 60 percent lower tumor volume. In terms of human equivalents, the 5 percent blueberry diet corresponds roughly to eating about two cups of fresh blueberries a day. In a second study, the blueberry-fed mice exhibited a reduced risk of the cancer spreading to other parts of their bodies.

What, then, would be a realistic headline to describe these results? How about "Large Daily Dose of Blueberry Powder May Reduce the Growth of a Rare Type of Artificially Induced

Breast Cancer in a Special Variety of Immune-Suppressed Mouse?" That wouldn't sell many papers, one would guess. And what do these mouse experiments mean for humans in terms of preventing or treating breast cancer? Not much. All we can do is mutter that blueberry extracts "warrant further investigation."

Marketers, of course, are not tethered to science. Any blueberry study that hints of some positive outcome, no matter how irrelevant it may be to humans, is enough to trigger an outburst of processed foods that feature blueberries on the packaging. You might think, for example, that Total, a cereal that loudly proclaims "blueberry" on the box, might actually contain blueberries. Well, you would be wrong. The "blueberries" inside are artificially colored and flavored bits of sugar mixed with fat. Even bagels or muffins that actually do have some blueberries contain insignificant amounts for any biological effect. Pure marketing hype.

How about the apple and pear study? Well, it really isn't a study about apples or pears. Researchers at Wageningen University in the Netherlands analyzed food frequency questionnaires filled out by some 20,000 people in terms of their fruit and vegetable consumption. Based upon the color of their "fleshy" portions, the fruits and vegetables were divided into green, orange and yellow, red and purple, and white.

The subjects were followed for ten years, a period during which 233 suffered a stroke. It turns out that stroke victims were more likely to have consumed fewer "white" fruits and vegetables than the other subjects. The researchers calculated that for every 25-gram increase in "white" fruit and vegetable consumption each day, the risk of stroke decreased by 9 percent. This may sound like a significant drop but really it is a small effect. Out of some 20,000 people, 233 suffered a stroke. That's roughly a 1.2 percent risk. A drop of 9 percent would mean the risk goes down to 1.1 percent. In other words, 1,000 people would have to increase their "white" intake by 25 grams to save one stroke.

So what are "white" fruits and veggies? Bananas, cauliflower, chicory, cucumber, pears, and apples. Within the "white" group, apples and pears were most commonly consumed, hence the catchy headline about apples and pears reducing the risk of strokes.

Now for a splash of critical thinking. First, food frequency questionnaires are notoriously unreliable. People have a hard time remembering what and how much they have eaten. And chances are that dietary habits change over the years. There is no guarantee that the pattern revealed by the questionnaire was followed over the ten-year follow-up period. Next, only white fruit and vegetable consumption was linked to a reduced incidence of stroke, not specifically apple or pear intake. Maybe the effect, if indeed there is one, is due to bananas or cauliflower.

This may sound like we're splitting fruits here. But there is a point. Some reports referred to the 9-percent decrease in strokes for every 25 grams of "white" fruits and vegetables and suggested that eating an apple a day (roughly 120 grams) can reduce the risk of a stroke by some 45 percent. That is some overly exuberant data-dredging; it's akin to inferring that blueberries can reduce the risk of breast cancer based on some mouse experiment.

While both the blueberry and apple studies are pretty hollow, they are at least in step with the plethora of publications that attest to the benefits of eating fruits and vegetables. So by all means fill up on blue, white, and whatever other colored fruits and vegetables you can find. But don't swallow the next headline about some "superfood" saving you from the clutches of the Grim Reaper. Exercise some critical thinking. And get some exercise. That can really reduce the risk of disease.

An Antidote to the Poisonous Tomato Legend

I ate my first tomato when I was about twelve years old. Actually, it wasn't even a tomato; it was tomato sauce on a pizza. I really don't know why, but growing up, I had a real aversion to tomatoes as well as to any food that contained them. And then a friend convinced me to try pizza. All of a sudden, a whole world opened up! Tomatoes, I discovered, taste great! And, as I would eventually learn, they were pretty healthy to boot. Perhaps this is why now, in my public lectures on food, I like to tell the tale of Robert Gibbon Johnson's tomato escapade.

As the story goes, in Salem, New Jersey, back in 1820, Colonel Robert Gibbon Johnson took out an ad in local newspapers inviting the public to gather in front of the courthouse on a Sunday afternoon to experience an epic event. He promised that, in full view of all, he would eat a tomato! At the time, tomatoes were thought to be poisonous and people were drawn by the possibility of seeing someone do harm to himself. Johnson, though, knew better, and thought Americans were depriving themselves of a delicious fruit. (Yes, the tomato, being the seed-bearing part of a plant, is a fruit.)

Colonel Johnson hired a little band to play a funeral dirge in the background as he picked up a tomato and took a large bite. Perhaps to the disappointment of the crowd, he didn't clutch his chest, foam at the mouth, or drop to the ground. He survived, and on that day the tomato industry was born, and we are all better for it. At least, so goes the oft-repeated story. Alas, repetition does not make a story true, no matter how compelling it may be.

Johnson really did exist. He was an elected member of the New Jersey State Assembly and at one time served as president of the Salem Horticultural Society. But in none of his writings did he ever make any mention of the tomato-eating incident. Indeed, we don't hear of the supposed epic moment until 1937, when it is highlighted in Joseph Sickler's book, *The History of Salem County*. It seems Salem's history needed a bit of spicing up and the "toxic tomato" story happened to fit the bill. But the truth is that by 1820, tomatoes, which are actually native to America, were regularly eaten. Long before Johnson's mythical, foolhardy experiment, President Thomas Jefferson had grown tomatoes in his own garden.

Myths are often born out of smidgens of facts. Sickler's "poisonous tomato" account may have been triggered by the close similarity of the tomato plant to others of the nightshade family, such as belladonna, which truly are poisonous. Nightshade plants contain potential poisons such as atropine in the case of belladonna, solanine in green potatoes, and tomatine in tomatoes. But of course, in toxicology, dosage is critical. Given that in a large enough dose it can cause harm, technically tomatine can be considered a poison, but the amount present in a tomato is negligible. In any case, the compound is found mostly in the flowers and leaves of the plant, which are not eaten. Tomatoes are not toxic, and the fact is that there is no evidence that anyone ever thought they were.

There are current myths about tomatoes as well. A popular one contends that some have been genetically engineered to keep them from freezing by inserting a gene from a fish known as the Arctic flounder. There are cartoons galore depicting tomatoes with fins or fish that have tomatoes for heads. While a genetically engineered tomato known as the FlavrSavr was briefly marketed with claims of improved taste (which it did not have), there are now no genetically engineered tomatoes on the market. Some researchers have indeed played with the possibility of inserting a fish gene that codes for an "antifreeze" protein into tomato plants to keep the fruit from freezing in case of a sudden cold spell, but this research has not borne fruit. In any case, this would not make the tomato into a fish-fruit hybrid. Fish have over 30,000 genes and no one gene makes a fish a fish. A tomato with one fish gene would still be a tomato. There could, however, be an issue with allergies, and if this project were ever to be commercialized, extensive testing would have to be done to ensure that someone with a fish allergy would not react to the engineered tomato.

A more realistic controversy about tomatoes focuses on whether organically grown tomatoes are in any way superior to conventionally grown ones. Like other fruits and vegetables, tomatoes contain compounds with antioxidant properties. While popular books and magazines loudly tout the almost-magical health benefits of antioxidants, the scientific literature is less compelling. There is no doubt that eating large amounts of fruits and vegetables is beneficial, but the exact reason is not clear. Produce contains thousands of compounds with biological properties, and specifically which of these compounds, if any, are responsible for the benefits is not known. Polyphenols, though, are reasonable candidates.

A recent Spanish study has shown that organic tomatoes have a slightly higher polyphenol content than those grown by conventional means, which is not surprising. Plants do not produce polyphenols for the benefit of humans. They produce the chemicals to help them survive stressful conditions such as a lack of nutrients in the soil or attack by insects and fungi. Without the application of synthetic pesticides and fertilizers, tomato plants are more stressed and produce more polyphenols. But this is unlikely to have any practical implications. In the context of an overall diet, the small differences between the polyphenol content of organic and conventionally grown produce is irrelevant.

Implying that organic tomatoes are better for us because they may have a slightly higher polyphenol content, especially in the face of a lack of studies showing that increased polyphenol content leads to better health, has about as much merit as the story about Robert Gibbon Johnson surviving the consumption of supposedly toxic tomatoes.

Leeches Then and Now

What did the jockey who never lost a race whisper into the horse's ear? "Roses are red, violets are blue, horses that lose are made into glue!" OK, so it's a groaner. But until the advent of polyvinyl acetate (PVA) and other synthetic glues in the twentieth century, the destiny of aging horses was indeed the glue factory. The collagen extracted from their hides, connective tissues, and hooves made for an ideal wood adhesive. Our word "collagen" for the group of proteins found in these tissues actually derives from the Greek "*kolla*" for "glue."

Not all aging horses were dispatched to the glue factory after their plow-pulling days came to an end. Some farmers found they could squeeze a little more profit out of the animals by assigning them another duty. They would become leech collectors! The elderly horses were driven into swampy waters only to emerge coated with the little bloodsucking worms. It seems the creatures found horses to be a particularly tasty treat! Since for many people suffering from various ailments the little parasites were just what the doctor ordered, the harvesting of leeches made for a lucrative business.

Leeches have actually been used in medicine since they were first introduced around 1500 B.C. by the Indian sage Sushruta, one of the founders of the Hindu system of traditional medicine known as "Ayurveda." That translates from the Sanskrit as "knowledge of life." Sushruta recommended that leeches be used for skin diseases and for various musculoskeletal pains. Ancient Egyptian doctors extended the indications, treating headaches, ear infections, and even hemorrhoids in this peculiar fashion. Galen, the famous Roman physician, used leeches to balance the four "humors," namely blood, phlegm, black bile, and yellow bile. Swollen, red skin, for example, was thought to be due to too much blood in the body and the answer was to have leeches slurp the excess.

Curiously, despite having no evidence for efficacy, bloodletting, either with leeches or by making an incision with a "lancet," became part of standard medical practice for more than 2,500 years! Monks, priests, and barbers got into the act along with physicians. In 1799, George Washington had more than half his blood drained in ten hours, certainly hastening his demise.

Many British doctors preferred leeches, especially in areas around the mouth, ears, and eyes, where lancing was a tricky procedure. They even learned how to encourage a leech to bite by stimulating its appetite with sugar or alcohol. But the creatures were in short supply, and had to be imported by the millions from France, Germany, Poland, and Australia, where they were often caught in nets using liver as bait. Sometimes poor children earned a little extra money by wading into infested waters to emerge, like the horses, with leeches attached to their legs. A gentle tug or a pass with a flame then relaxed the bloodsucker's grip before much damage ensued. Good thing, because leeches can be pretty nasty once they latch on. Remember Humphrey Bogart flailing about in *The African Queen* while trying to rid himself of the little vampires?

The lack of leeches caused some physicians to explore recycling techniques. A single leech usually becomes satiated after filling up on about 15 milliliters of blood and then falls off. But then if it is plunked into salt water, it will disgorge the blood and is soon ready for another round. A German physician even developed a technique to encourage continued sucking by making an incision in the leech's abdomen, allowing for the ingested blood to drain out as fast as it went in. It seems the leech wasn't much bothered by this affront to its belly and would go on sucking for hours. Amazingly, leeches were sometimes used internally. To treat swollen tonsils, a leech with a silk thread passed through its body would be lowered down the throat and withdrawn when it had finished its meal. Sometimes the creatures were even introduced into the vagina to treat various "female complaints." The literature is vague about how this was done but one account suggests that the technique required a clever nurse.

While bloodletting as a general treatment for ailments has been drained out of the modern medicine chest, there is still work for leeches. That's because their saliva is a complex chemical mix of painkillers and anticoagulants. Hirudin, for example, is the protein that keeps the blood flowing steadily after the initial bite is made, and is so effective that the blood will not coagulate for quite some time even after the leech falls off. Indeed, these bloodsucking aquatic worms have received approval from the U.S. Food and Drug Agency as a "medical device."

Surgeons have been known to use leeches after reattaching ears, eyelids, or fingers that have been severed, as well as after skin grafts. This has to do with the fact that arteries are easy to reconnect but veins are not. Eventually new capillaries do form to reconnect veins, but in the meantime the finger or ear fills with blood, which then clots and causes problems with circulation. A leech will drain the excess blood at just the right rate and can prevent blood clot formation by injecting hirudin. This is such a potent anticoagulant that it holds hope for dissolving blood clots after a heart attack or stroke. Unfortunately, hirudin is too difficult to extract from leeches, but it can potentially be produced through genetic engineering techniques.

Where do physicians get leeches today? No need for horses. They can order them directly from the French firm Ricarimpex. One would think that, after helping to save a finger or an ear, the useful little critters would be rewarded. But their destiny is death in a bucket of bleach. Not any better than ending up in a glue factory.

Crying Wolf

California. Home to glitzy movie stars, aging hippies, outstanding wines, Silicon Valley, and legislators who are remarkably knowledgeable about toxicology. They must be, because just about everywhere you go in the state, be it a restaurant, supermarket, gas station, amusement park, or airport, you're confronted with signs and labels that warn about the presence of chemicals "known to the State of California to cause cancer, birth defects, or reproductive harm." Surely these lawmakers must have special insight into the health effects of the plethora of natural and synthetic chemicals to which we are exposed on a daily basis, because no other regulatory agency anywhere has deemed it necessary to provide such warnings. But since manufacturers are not keen to produce products with different labels for different markets, we all get treated to labels that threaten us with the prospect of cancer should we handle items such as vacuum cleaners or Tiffany-style lamps.

The ubiquitous labels began to appear in 1986 after the passage of "Proposition 65," a law that prohibited businesses from knowingly exposing individuals to chemicals that cause cancer or developmental problems unless they provided a clear and reasonable warning. Whether that warning is actually reasonable is debatable. A list of the chemicals in question was compiled by California's Office of Environmental Health Hazard Assessment and has since expanded to over 800 substances. Basically, Proposition 65 was an attempt to put the "precautionary principle" into practice, the idea being to avoid a chemical that might present a risk even if the evidence is not absolute. Sounds like motherhood and apple pie. But according to Proposition 65, even motherhood and apple pie would be suspect, since progesterone, the "pregnancy hormone," as well as acrylamide and formaldehyde, both present in apple pie, appear on California's list.

How does any substance wind up as a member of this rogues' gallery? Evidence for human carcinogenicity is not the criterion! If a chemical causes any sort of cancer or developmental problem in any animal, at any dose, it qualifies. So alcohol, tobacco, benzene, benzopyrene, lead, and arsenic, all proven human carcinogens, are of course on the list, but so are aspirin, nickel, phthalate plasticizers, laughing gas, mineral oil, and phenolphthalein, the acid-base indicator that is a staple in high school chemistry labs. The latter group has been flagged based on animal studies that may have little relevance to humans. Given that testosterone, progesterone, and estrogen, all naturally occurring hormones, are on the list, it's a wonder that people aren't made to walk around with labels on their forehead declaring the presence of chemicals known to the state of California to be dangerous. I guess we are exempt because we don't devour each other. On the other hand, we don't eat vacuum cleaner hoses, electrical wires, or the plastic panels that cover airline counters, either.

Yes, airline counters. Last year, when approaching an agent at Los Angeles International, I was stunned to see a huge warning sign on the front panel of the desk. I thought it must be some sort of notice alerting me to the dreadful consequences of trying to smuggle nail clippers or perhaps more than 100 milliliters of shampoo aboard. But no! I was being warned, "this area contains chemicals known to the State of California to cause cancer." Since I had no great desire to take a bite out of the counter (although it may have been an improvement over airline food), I was not concerned about any health risk, but the existence of this warning was disturbing. How were people supposed to react to it? Hold their breath and flee in panic? Ridiculous.

On to the vacuum-cleaner issue. Why the warning? Because the polyvinyl chloride (PVC) insulation on the electrical wire contains traces of lead. PVC is an excellent insulator, but it has to be molded into its final shape with heat. Therein lies a problem. High temperatures lead to the decomposition of the PVC, which can be countered by the addition of small amounts of lead compounds. But since the electrical wires are handled when the vacuum cleaner is plugged in, trace amounts of the lead might in theory leach out from the wire, and if the hands are not washed, a trace of the trace might end up on any food that is handled, and a trace of the trace of the trace might be ingested. The chance of such exposure having any effect on health is remote, but what the warning may do is scare some people away from using a vacuum cleaner. And inhaling dust is not exactly conducive to health.

Proposition 65 also has a "bounty hunter" provision that encourages individuals or groups to sue if a warning label is absent. Some of these suits were judged to be frivolous, such as the one against chocolate manufacturers for not warning about the presence of naturally occurring, but biologically irrelevant, traces of cadmium in their products. That was quickly dismissed, but there have been more than 2,100 other lawsuits resulting in millions paid out in settlements and more than a billion dollars spent by industry to reformulate products to prevent having to put on warning labels. Who has benefited the most? Probably the lawyers.

Proponents of Proposition 65 argue that the reformulated products are safer. Is there any evidence for this? The law has been in effect for over twenty-five years, but cancer incidence in California is basically the same as elsewhere in the U.S., except for smoking-related cancers, which have decreased. The state certainly does deserve credit for its strong anti-smoking stand, but no lives have been saved by frightening people away from vacuum cleaners or airline counters or washing machines or garden hoses or flashlights. Why? Because Proposition 65 is not based on exposure, it is based on content! It makes no sense to have a warning just because a substance that causes some problem at a high dose in test animals is present in a product in amounts that are so trivial that no other regulatory agency deems them to be a risk. On the other hand, the law itself may present a risk. Crying about phantom wolves stresses people needlessly, and stress is a risk factor for disease. In any case, the omnipresent warnings become so much a part of the landscape that they are usually ignored. That may also be the fate of warnings when a real wolf comes to the door.

What's for Dinner?

So, what should we have for dinner? It seems a simple question. But is it ever difficult to answer! Unfortunately, tasty and healthy don't always coincide. And just what is "healthy," anyway?

For close to forty years, I've pored through countless research papers and media accounts about food and nutrition. I've interviewed some of the world's top researchers in this area. My shelves sag with the weight of dozens and dozens of books on the subject and I've even written one myself that deals exclusively with food, as well as thirteen others with plenty of nutritional connections. And after all that, I'm still mystified about what we should have for dinner. But not completely. The wheat is slowly being separated from the chaff.

One thing is for sure: there's no shortage of nutritional information or of opinions about what we should eat. Dr. Robert Lustig, a pediatric neuroendocrinologist at the University of California, believes that many health problems, obesity in particular, can be traced to consuming too much fructose. His video on the subject has gone "viral." University of Missouri Professor Frederick vom Saal is of the opinion that obesity can be linked to bisphenol A, a chemical that can leach from the lining of canned foods. Dermatologist Dr. Robert Bibb, in his book *Deadly Dairy Deception*, makes a case for dairy products being the cause of prostate and breast cancer. Dr. Neal Barnard, president of the Physicians' Committee for Responsible Medicine, goes even further in *Eat Right, Live Longer*, claiming that salvation lies in avoiding all animal products.

Cardiologist Dr. William Davis sees no problem with meat, but sees wheat as the real bogeyman. According to him, the grain's polypeptides cross the blood–brain barrier and interact with opiate receptors to induce a mild euphoria that in turn causes addiction to wheat. As he describes in his best-selling book *Wheat Belly*, this results in fluctuating blood sugar levels that in turn create hunger and lead to obesity as well as numerous other health problems. How does Davis know all this? Apparently, his patients lose weight on a wheat-free diet and recover from all sorts of diseases. Has he published any of this in peer-reviewed journals? Not that I can find.

Davis would probably find a kindred spirit in Dr. Drew Ramsey, a clinical professor of psychiatry at Columbia University in New York. His book is called *The Happiness Diet: A Nutritional Prescription for a Sharp Brain, Balanced Mood, and Lean, Energized Body*. What is that wondrous prescription? It seems simple enough. If you want to be happy, stay away from bagels. According to Dr. Ramsey, "At first bagels boost a person's energy, but after a few hours you come crashing down looking for another fix in the modern American diet. That crash can cause people to feel irritable, lightheaded, or sad." Really? Maybe if they eat American bagels. I think legions of happy Montreal bagel lovers would disagree.

Journalist Gary Taubes maintains that not only wheat, but all carbohydrates, should be limited. In *Good Calories, Bad Calories*, he has gathered a massive amount of information to "prove" that excessive consumption of carbohydrates is the cause of heart disease, cancer, Alzheimer's disease, and type-2 diabetes. He advises against a low-fat diet. Dr. Dean Ornish, in *Eat More, Weigh Less*, would take issue with Taubes. He puts his cardiac patients on an extremely low-fat, high-complex-carbohydrate diet and has evidence that deposits in arteries actually regress.

I could go on and on about all the dietary advice that floods us. In *Soy Smart Health*, Dr. Neil Solomon claims that eating soy can decrease the risk of breast cancer, heart disease, and osteoporosis, while in *The Whole Soy Story*, Dr. Kaayla Daniel links soy to malnutrition, digestive problems, thyroid dysfunction, cognitive decline, reproductive disorders, heart disease, and cancer. Go figure.

Other authors suggest that our health is being undermined by monosodium glutamate (MSG) or artificial sweeteners or trans fats or pesticide residues or cooking in Teflon pans or genetically modified organisms or chlorinated water or acrylamide or phthalates or hormone residues or . . . or . . . or . . . And there is no lack of advice about how to get our health back on track. All we have to do is drink some esoteric juice, pop some sort of dietary supplement, gorge on some superfood, or eat like the Greeks or Chinese. So who do we listen to? The "experts" can't all be right.

When I throw all the divergent opinions into my mental flask and distill the essence, I come up with something like *The Okinawa Diet Plan*. This fascinating and very well-researched book chronicles the lifestyle habits of the longest-lived population in the world. We're not looking at some mythical Shangri-La here. In the Japanese islands of Okinawa, we have our sights on a people whose unusual longevity and good health is well documented. So is the fact that Okinawans do not gain significant weight as they age! Why? Because they consume 1,600 calories a day, at least 500 less than we do. And they do this while eating half a pound more food. It's all a matter of what sort of food: no hamburgers, hot dogs, or smoked meat here. And no soda pop. But they do eat plenty of food with very few calories per gram.

The lower the calorie density, the more food can be eaten without gaining weight. Basically this means a plant-based diet. For example, broccoli, mushrooms, and carrots check in at about 0.4 calories per gram, tofu at 0.7, bread or meat, of which Okinawans eat very little, at about 3.0, and oils weigh in at 8.8. Michael Pollan, in his popular book *The Omnivore's Dilemma*, echoes the Okinawan way of eating: "eat food, not too much, mostly plants."

I think the fog of confusion about nutrition is finally starting to clear. So, we can take another shot at the "what's for dinner?"

quandary. I think we'll go with an asparagus tofu stir-fry and a hearty bean stroganoff. And tomorrow? My good old vegetarian goulash. All the while keeping in mind the Okinawan philosophy of "hara hachi bu," which translates as "stop filling your stomach when you're 80 percent full." But I still won't mothball my barbecue. After all, there's more to life than worrying about whether every morsel we put in our mouth is "healthy" or not.

Twinkies, M&Ms, and Weight Loss

The laws of thermodynamics will never be repealed. To lose weight, you have to expend more calories than you take in. And as far as shedding pounds goes, it doesn't much matter how you cut calories. That's precisely the point Kansas State University nutrition professor Mark Haub wanted to make to his class a couple of years ago. He figured that the prospect of a professor becoming a guinea pig was sure to get students' attention, especially when there was a chance that he might regret the idea. And it seemed that going on a "Twinkie diet" was a pretty good candidate for a "gee, I shouldn't have done that" type of experience.

Dr. Haub's goal was to demonstrate that weight loss is just a matter of calories in and calories out. Burn more than you take in, and the weight will drop. You can eat nothing but pure sugar and still lose weight, as long as the calorie count is less than what is normally needed to fuel the body. Haub estimated that a reduction in his intake from the usual 2,500 calories to about 1,800 would result in the loss of one to two kilograms, or approximately 4.4 pounds, per week. Eating only sugar seemed a little extreme, so he decided on a mix of Twinkies,

Little Debbies, Ho-Hos, Doritos, and Oreos, a diet that would send a shiver up any nutritionist's spine. But he hoped to show that even with this dietary nightmare of sugar and fat he would drop pounds. Of course, Dr. Haub knew that the body also has requirements for protein, vitamins, and minerals, so he did include a few low-calorie vegetables, some milk, and a multivitamin-mineral supplement in his regimen.

After four weeks, it was already apparent that the battle against bulge was being won. Amused students began to prattle about their prof's crazy antics, and before long the press got wind of the Twinkie diet. Initially Haub's "convenience store diet" was to last only a month, but with the increased media interest, the previously portly prof decided to extend the experiment until he reached normal weight. In just ten weeks, he shed 12 kilos, dropping his body mass index from 28.8 to a normal 24.9. Impressive. Point made. You can lose weight even on a nightmarish diet, as long you keep the calorie count down. But at what cost?

Surely a regimen based on fat and sugar must be detrimental in terms of heart disease! It must send cholesterol and triglyceride levels soaring. Well . . . nope! Haub's "bad cholesterol," or LDL, dropped by 20 percent, his "good cholesterol," or HDL, went up by the same amount, and his triglycerides decreased by almost 40 percent. Actually, this isn't all that surprising. Weight loss, no matter how it is achieved, is known to improve these markers of cardiovascular risk. But of course there's more to health than cardiovascular risk. The long-term consequences of a diet so low in fruits and vegetables might well include an increased risk of cancer.

So what happened to Dr. Haub's weight after he went off his Twinkie diet? After all, experience tells us that losing weight is not all that hard, but keeping it off is murder. The professor is actually doing OK. He says the experience has taught him to focus on portion size. He's eating about 2,200 calories a day, his cholesterol is up slightly, and he has gained back a couple of pounds. But Haub eats everything, and still munches on the occasional "snack cake," though he had to give up Twinkies for a few months when the Hostess company filed for bankruptcy. But to many people's delight and nutritionists' horror, the brand was purchased by another company and Twinkies are now back on the market.

Admittedly, an unpublished "study" with a single subject as its subject doesn't prove that the composition of the diet makes no difference in terms of weight loss. But how about a Harvard study published in the *American Journal of Clinical Nutrition*? This one had 811 participants who followed one of four different diets for two years. Each diet furnished roughly 750 calories less than normal energy requirements but they differed dramatically in their protein, fat, and carbohydrate composition.

After two years, average weight loss was about 4 kilograms (8.8 pounds), regardless of which diet was followed. Clearly, it's a case of calories in and calories out. So it really should come as no surprise that any diet, be it a cookie diet, a muffin diet, a cabbage soup diet, or a cricket, cockroach, and flea-leg diet will result in weight loss, as long as the total calories provided are less than the body's needs. It's always the same story. People adopt a diet, lose weight thanks to some novel attention-grabbing gimmick that restricts caloric intake, and then balloon again because the gimmicky diet cannot be sustained. Strangely, it doesn't stop them from imagining that the next diet that comes along is the "one."

And maybe next time, all they will have to do is imagine. Researchers at Carnegie Mellon University in Pittsburgh asked subjects to imagine one of three scenarios. One group was told to think of inserting thirty-three coins into a laundry machine, another was asked to imagine inserting thirty coins and eating three M&M candies, and the last group was to think about inserting three coins and eating thirty M&Ms. The results were reported in the prestigious journal *Science*.

After visualizing dropping coins into machines and candies into mouths, the participants were presented with bowls of M&Ms and allowed to indulge to their heart's delight. Subjects who had imagined eating thirty candies consumed significantly fewer than the ones who had concentrated on inserting coins into laundry machines. There was no word on whether the coin inserters were less motivated to do laundry.

So is this enough evidence to resuscitate the Twinkie diet? Maybe all you need is a little imagination to lose weight. To cut down on calories, just visualize snacking on some Twinkies before digging into your next meal! Seems safe enough. I can't imagine that just thinking about junk food is risky. But obesity sure is. I would suggest, though, that the best bet for success with the "imagination diet" is to do the imagining while running on a treadmill.

The Rise, Fall, and Possible Rise of Tropical Oils

"Is your company an accessory in the deaths of untold numbers of heart attack victims?" So began a letter sent to some 11,000 food-industry executives back in 1985. The saturated fats they were pumping into their processed foods, the accusation went, were killing Americans. Companies were urged to purge their products of lard and beef tallow, as well as of "tropical oils" derived from palm or coconut. The letter did not come from any sort of regulatory agency, it came from an individual. A very wealthy individual.

By age fourty-four, Phil Sokolof had made a lot of money in the construction supply business. He was lean, didn't smoke, didn't have high blood pressure, and exercised regularly. The last thing on his mind was a heart attack. But it struck and nearly killed him. The cause, he concluded, was that he had been a "student in the greasy hamburger school of nutrition." Saturated fats in his diet had done him in and they were doing the same to others. They had to go.

When his letter provoked no more than a yawn, Sokolof took action. He put his money where his mouth was. The mouth that had gorged on all that greasy food would now disgorge the anti–saturated fat message. He spent millions on full-page ads in major newspapers across the country with the headline "The Poisoning of America!" "We implore you," Mr. Sokolof said in the ad, "do not buy products containing coconut oil or palm oil. Your life may be at stake." The ads listed quantities of saturated fats in various brands of cookies, cakes, cereals, and fast foods and portrayed major food companies such as Kellogg's, Nabisco, and McDonald's as the demons who were sending Americans to an early grave.

The public outcry was huge and it dealt a real financial blow to a number of food companies. They responded with commitments to phase out the devilish tropical oils and replace them with unsaturated vegetable fats. After all, these were not burdened with the baggage of causing heart disease by raising levels of cholesterol in the blood. The soybean oil industry rejoiced. But there was a problem.

Palm oil and coconut oil are solid fats that have excellent stability. They also give just the right texture and taste to processed foods. Unsaturated vegetable oils, on the other hand, react with oxygen readily and consequently do not keep as well. Also, they are liquids, not particularly suitable for the production of baked goods. There is, however, a remedy for this situation. Unsaturated fats can be treated with hydrogen, and the resulting partially hydrogenated vegetable oils can serve as effective substitutes for saturated fats. The industry's drive to switch from saturated tropical oils to partially hydrogenated vegetable fats was praised by Sokolof.

It wasn't long before food companies discovered that they had gone from the frying pan into the fire. The hydrogenation process, they learned, produced some side products. These were the nefarious trans-fatty acids! They were as guilty as saturated fats, if not more so, of increasing artery clogging low-density lipoprotein (LDL) cholesterol and decreasing the artery-protecting high density (HDL) version. Trans fats became persona non grata and public officials stumbled over each other in their rush to introduce legislation to ban them. And guess what is re-emerging as a substitute. Yes, those maligned tropical oils!

Actually, it turns out that not all saturated fats are equally wicked. The number of carbon atoms in their molecular structure determines the extent to which they affect blood cholesterol levels. Coconut oil is composed mostly of medium-chain fatty acids. Having only six to twelve carbons in their structure, as opposed to eighteen in most animal fats, they may not be as damaging as once thought. While medium-chain fatty acids do raise "bad cholesterol," they also raise the "good" kind, and can result in a more favorable ratio of good to bad.

Still, the current consensus is that replacing saturated fats and trans fats in the diet with polyunsaturated fats is the wise thing to do. But with the amount of publicity that the diet–heart disease link has received, consumers are likely to come away with the message that a meeting with the undertaker is just around the corner unless they replace the evil saturated fats with the angelic unsaturated ones. While there is little doubt that such a substitution decreases the risk of heart disease, the impact is not as dramatic as is commonly believed.

A recent thorough analysis compiled all the published studies in which subjects had replaced saturated fats in the diet with unsaturated ones over an average follow-up period of about five years. With a major effort at making the substitution, the risk of heart disease was reduced by roughly 10 to 15 percent. That sounds pretty significant, but it is somewhat misleading. Reporting results in terms of percentage often is.

It is instructive to look at the actual numbers. There were approximately thirteen thousand subjects in the combined trials, divided into control groups and intervention groups. In the control group, where no dietary modifications were made, there were 8.3 combined heart attacks and deaths for every 100 subjects. In the intervention group, where polyunsaturates were substituted for saturates, there were 7 events for every 100 subjects. The difference between 8.3 and 7 is your 10 percent difference.

But another way to look at this is that reducing saturated fats in favor of unsaturated ones saves about one cardiac "event" for every hundred people who make the effort. That doesn't sound as impressive as a 10-percent reduction in risk. It isn't negligible, though, especially if you are the one being saved. And when whole populations are considered, the benefits may be considerable.

Phil Sokolof was without a doubt well-meaning, and he deserves credit for bringing public attention to the issue of fats in the diet. Even back in the 1980s, though, his evangelistic, overzealous crusade against tropical oils had less scientific support than he claimed. For one, tropical oils were minor contributors to saturated fat in the diet, providing only about 2 percent of total fat. And while there was evidence that saturated fats in animal foods were capable of raising cholesterol and increasing the risk of heart disease, there was no evidence that tropical oils did the same. They were implicated because they were also "saturated," even though other features of their molecular structure were quite different.

Tropical oils are not demons; no single food components are. The quality of the diet always comes down to an analysis of everything that is consumed. Interestingly, coconut oil is now being investigated as a possible treatment for Alzheimer's disease.

Thinking About Coconut Oil

Think about this: what has no mass, doesn't occupy space, has no mobility, cannot be touched, and yet exists? A thought! And what a mysterious thing it is! Just about all we know for sure is that it is created in the brain and that there is an energy requirement to generate it.

Whenever we think, the brain "burns" more glucose, which is its main fuel supply. It stands to reason that any sort of inhibition of this glucose metabolism can have a profound effect on brain function. We know, for example, that a rapid drop in blood glucose, as can be precipitated by an overdose of insulin, quickly causes a deterioration in cognitive performance. This is because so much glucose is absorbed by muscle cells that little is left for the brain.

Alzheimer's disease is characterized by a progressive decline in the rate of glucose metabolism in the brain. This impaired use of glucose is paralleled by a decline in scores on cognitive tests. Exactly why glucose use is affected in Alzheimer's is not clear. It may be a function of the buildup of amyloid protein deposits that are the hallmark of the disease, although it is also possible that the deposits are not the cause, but are rather the result, of impaired metabolism. In any case, improving the brain's ability to generate energy in the face of low glucose metabolism seems a worthy avenue to explore.

The most obvious approach would be to supplement the diet with glucose and provide sufficient insulin for its absorption into cells. But insulin cannot easily be delivered specifically to the brain, and its systemic administration can cause problems in other tissues. So is there another option? A clue can be found in studies of people who are experiencing starvation. When there is a lack of glucose available from the diet, the body tries to meet the brain's demand for energy by tapping its abundant stores of body fat.

Fat, however, cannot be used directly as fuel; it first has to be converted to smaller molecules called ketone bodies. The buildup of these in the bloodstream results in ketosis, a condition that is not encountered when there is an adequate intake of carbohydrates, the source of glucose. It can, however, occur in diabetes when an insulin shortage prevents glucose absorption into cells, which then have to resort to the use of ketone bodies to supply energy. That's why acetone, a ketone body, appears in the breath of diabetics who fail to administer their insulin properly. Ketosis can also be encountered when low carbohydrate regimens such as the Atkins diet are followed. It is the breakdown of fat to yield ketone bodies that results in weight loss.

Now back to Alzheimer's disease. An extremely low-carbohydrate diet can conceivably increase ketone bodies delivered to the brain, but such diets are difficult to follow and may not be healthy for other reasons. But there may be another approach. It turns out that not all forms of dietary fat are handled by the body the same way. So-called long-chain fatty acids, composed of at least twelve carbons, as found mostly in animal products, are readily stored by the body, whereas the "medium-chain fatty acids" that contain six to twelve carbons tend to be metabolized in the liver to ketone bodies. This presents a potential therapeutic application for Alzheimer's disease. Why not just supplement the diet with medium-chain fatty acids? They're not hard to find. You don't have to look further than coconut oil.

At least one published trial lends support to the idea. Patients in the early stages of Alzheimer's disease showed an improvement in cognitive performance tests administered ninety minutes after treatment with a single forty-gram dose of medium-chain fatty acids. This finding flew pretty well under the public radar until pediatrician Dr. Mary Newport's story started to circulate on the internet. Actually, it was her husband's story that got people talking. Steve Newport was diagnosed with Alzheimer's and was fading quickly. His wife did what most people do these days: she let her fingers do the walking on the keyboard. As a physician, she knew that the drug treatments available were not very effective and became intrigued when she came across the research that had linked medium-chain fats to increased metabolism in brain cells. What was to lose by giving her husband a couple of tablespoonfuls of coconut oil every day?

The very next day, Steve was scheduled for a routine cognition test and showed a surprising improvement over his previous performances. Dr. Newport obviously decided to continue the regimen and reports that after two months her husband was once more reading avidly, resumed jogging, and even started to do volunteer work at a hospital. But should he miss his morning oil, he quickly becomes confused and experiences tremors. Swallowing the regular dose brings quick improvement.

So what are we to make of all this? Is there a cure for Alzheimer's disease that is being ignored by conventional medicine? Not likely. But that is not to say there's isn't something to the medium-chain fatty acid story. However, it is a little disturbing that the source for the internet buzz is an article written by Dr. Frank Shallenberger. Let's just say this good doctor is not a candidate for a staff position at Harvard Medical School. Following multiple disciplinary actions for gross incompetence, he surrendered his California license and moved to Nevada, where he later pleaded guilty to another count of medical malpractice. He now writes a newsletter about "real cures," such as "ozone therapy," and pushes medium-chain fatty acids for Alzheimer's disease.

What we have here is one interesting study published in the literature that in no way shows reversal of Alzheimer's, an intriguing personal account that begs for independent verification, and some overly optimistic statements from a physician who has had disciplinary actions against him for incompetence.

But let's not throw the baby out with the bathwater. The theory behind boosting levels of ketones such as acetone to enhance cellular energy production in the brain has merit and needs further exploration. Don't think, though, that drinking nail polish remover is the way to go. But coconut oil? Well, let's hold that thought.

Free Radicals Bad, Antioxidants Good: Is That So?

There is one thing we know for sure about antioxidants: they sell products. Unfortunately, that is just about the only thing we know for sure about this fascinating class of chemicals. In the public mind, though, antioxidants are superheroes that wage war against those evil free radicals that conspire to rob us of our health and our youth. And according to a variety of supplement promoters, the antioxidants that are naturally present in our food supply are not enough to protect us from the free-radical onslaught. We need reinforcements in the form of whatever pill, capsule, or potion they happen to be pushing. And the concoctions usually come with plenty of testimonials about lives turned around. But what they lack is compelling evidence.

The story usually goes something like this: we need oxygen to live. No contesting that. Any student who has studied glycolysis and the dreaded Krebs cycle will recall the critical role that oxygen plays in the production of cellular energy. Basically, glucose reacts with oxygen to yield carbon dioxide, water, and energy. But there are also some byproducts. These are the notorious free radicals, also referred to as reactive oxygen species, or ROS. And they are reactive. Should they take aim at important biomolecules such as proteins, fats, or nucleic acids, they can wreak molecular havoc.

Our body, however, doesn't just stand by as the electron-deficient free radicals try to satisfy their hunger for electrons by ripping essential molecules apart. It musters its defenses. And those defenses are the antioxidants, a wide array of compounds linked by the ability to neutralize reactive oxygen species. They encompass enzymes such as superoxide dismutase, vitamins A, C, and E, and various polyphenols derived from plant products in our diet. Since plants produce oxygen through photosynthesis, they have had to evolve protective mechanisms to deal with oxidation. It is reasonable to propose that we can benefit from the antioxidants they churn out. So far, so good.

Populations that consume more fruits and vegetables are healthier. That is also more or less correct. But why should this be so? This is where the fly plunges into the ointment. The seductive argument is that produce is loaded with antioxidants and that by scavenging free radicals these chemicals are responsible for the health benefits. But fruits and vegetables are also loaded with all sorts of compounds that have biological activity unrelated to free-radical scavenging. Flavonols in cocoa beans, for example, dilate blood vessels by triggering the formation of the messenger molecule nitric oxide. Isoflavones in soy interact with estrogen receptors. Curcumin in turmeric inhibits an enzyme that catalyzes the formation of pro-inflammatory prostaglandins. Salicylic acid in apples has an anticoagulant effect. In spite of all this, the focus has been on antioxidants. A simple formula emerged: free radicals are bad, antioxidants are good. And the marketers ran with that one. Fast enough to blur the facts.

Any substance that showed antioxidant potential in the laboratory was elevated to the status of a quasi-drug. Shelves sagged under the weight of exotic juices, green tea pills, pine bark extracts, capsules filled with various carotenoids, and, of course, vitamins C and E in every conceivable form. The contest was on for the antioxidant championship of the world. Products vied with each other to claim the highest oxygen radical absorption capacity, or ORAC rating. ORAC is a measure of the ability of a sample to neutralize free radicals in a test tube. But the body is not a giant test tube, and ORAC values do not necessarily translate into biological significance. Many polyphenols that can put on an impressive antioxidant performance in the test tube may not even be absorbed from the digestive tract.

Vitamins C and E, along with beta-carotene, readily decimated free radicals in lab experiments and became the poster boys for antioxidant supplements. But when researchers got around to carrying out clinical trials, the results were disappointing. Most found no benefit. One actually showed an increase in lung-cancer risk in smokers taking beta-carotene supplements. Some studies even claimed an increased risk of premature mortality in people who regularly supplemented with antioxidants. What's going on here? Could it be that free radicals are not the villains they have been made out to be? As is so often the case in science, issues that seem straightforward on the surface become more complicated with a little digging. So let's dig a little.

It turns out that white blood cells generate and unleash free radicals in their fight against bacteria and viruses. So clearly, in the right quantities, at the right time, free radicals can be health enhancing. Furthermore, the production of free radicals for such defense purposes is sensed by other cells that then fire up their internal defenses and produce enzymes, such as catalase and superoxide dismutase, that can deal with larger numbers of potentially dangerous free radicals. Sort of like how exposure to a small amount of toxin prompts the system to deal with larger insults.

One interesting theory suggests that antioxidants in food may actually work by generating small doses of health-promoting free radicals. When an antioxidant neutralizes a free radical by donating an electron, it itself becomes a free radical, but a potentially much less damaging one. Still, it may be threatening enough to stimulate the body's own antioxidant defenses. Antioxidant supplements may not work as well as antioxidants found in food because the doses are too high, and they may suppress free radical formation excessively. Sounds far-fetched? Well, consider this.

It is well known that exercise improves insulin sensitivity, which in turn helps manage type 2 diabetes. However, exercise also increases the formation of reactive oxygen species as cells "burn" more glucose to generate the energy needed. In a German study, forty healthy young men were given an exercise regimen to follow for four weeks. Half the subjects were asked to take a daily supplement of 1,000 milligrams of vitamin C and 400 IU of vitamin E. Surprisingly, insulin sensitivity improved only in the men not taking the supplements! Furthermore, production of superoxide dismutase and glutathione peroxidase, the enzymes that protect against free radicals, was increased by exercise, but again only in the subjects not taking the supplements. It seems that the free radicals produced by exercise-induced oxidative stress provide the signal for increased insulin sensitivity and for revving up antioxidant defenses. The researchers concluded, "supplementation with antioxidants may preclude these health-promoting effects of exercise in humans." So, I think I'll stick to getting my antioxidants from my daily five to seven servings of fruits and vegetables.

Life sure is complicated. There are no simple solutions. As H.L. Mencken said, "For every complex problem there is a solution that is simple, neat, and wrong." And the relationship between diet, health, and supplements is indeed a complex problem.

A Health and Education Act? Really?

Is it natural or is it synthetic? Paradoxically, that's the pertinent question when it comes to the legality of marketing dimethylamylamine (DMAA) as a dietary supplement in the U.S. And it all has to do with a piece of legislation passed in 1994 known as the Dietary Supplement Health and Education Act (DSHEA). This bizarre act has little to do with the relationship between dietary supplements and health, and nothing at all to do with education. It is a scientific travesty.

DSHEA was in large part the handiwork of Senator Orrin Hatch of Utah, a state populated by numerous supplement manufacturers who felt threatened by the Food and Drug Administration's attempts to combat the mushrooming cases of health fraud. The supplement industry went to work and whipped the public into a frenzy with their false claims that the government intended to take away their freedom to choose dietary supplements, including vitamins. This was utter nonsense, but the "freedom to choose" argument mustered enough traction to allow the passage of DSHEA by a Congress more interested in appeasing the electorate than in championing science.

DSHEA stated in somewhat vague terms that manufacturers and distributors were responsible for ensuring the safety of their products. There were two requirements under the act: any supposed active ingredients had to be found in nature, and if a novel ingredient was to be introduced, the Food and Drug Administration had to be notified, supported by "evidence of safety." The burden was on the FDA to prove that a product was unsafe if it wished to remove it from the market. As far as any proof of efficacy was concerned, there was absolutely no requirement. Useless but harmless supplements could be happily marketed. When DSHEA was passed, it had a proviso for substances that were already on the market in 1994. If they had a "history of safe use," which was undefined, they could continue to be sold without any further requirement.

It was under this wacky clause that ephedrine was grandfathered as a dietary supplement. Isolated from the ma huang plant, the drug had a long history of use in traditional Chinese Medicine as a treatment for respiratory problems. But nobody had actually monitored its safety. Since the drug can also boost metabolism, it was particularly attractive to marketers hoping to cash in on Americans' expanding waistlines. By 2000, however, the FDA had amassed sufficient evidence about heart disturbances to make ephedrine the first supplement to be banned under DSHEA. The ban also precipitated legislation requiring manufacturers to disclose any reports of adverse reactions they received.

Supplement manufacturers who had been cashing in on ephedra's reputation as a metabolism booster were hard hit. They looked for replacements and came up with DMAA, a compound that, like ephedrine, had a chemical similarity to amphetamine. It had first been synthesized and patented way back in 1944 by the Eli Lilly Company as a nasal decongestant, but never made it to the market. In 2006, Ergopharm, an American supplement company, resurrected DMAA and began to market it as a body-building and weight-loss supplement. To justify introduction under DSHEA, the company had to submit safety data and had to show that the compound occurred in nature. They managed to dredge up a single Chinese study that claimed DMAA was found in geranium extract but numerous attempts by other researchers to reproduce this analysis have failed. There is presently no substantive evidence that DMAA occurs in nature. Nor did the company submit any safety data to the FDA as is required for a novel substance.

The World Anti-Doping Agency was quick to claim that the compound is a stimulant and banned its use, as did Major League Baseball. However, the drug skittered under the public radar until accounts of elevated blood pressure, nervous system disorders, and psychiatric complications began to appear in the media. Finally, a possible link between the deaths of two American soldiers and the drug triggered a closer look by regulatory agencies. The soldiers, both of whom had been taking DMAA, suffered heart attacks during training. Although no link was proven, the supplement was pulled from all stores on military bases. Just how much DMAA the soldiers had taken isn't clear.

In any case, the FDA has now determined that conditions for the sale of DMAA have not been met and has accordingly sent warning letters to a number of fitness supplement companies advising them to stop selling DMAA or risk possible seizure of the products. There is evidence, the FDA has declared, that DMAA narrows blood vessels, possibly causing an elevation in blood pressure, which may lead to cardiovascular events including heart attack. Canada, with a different framework for regulations, has banned the drug outright, as have a number of other countries.

Needless to say, the actions by the American and Canadian regulatory agencies have not been without opposition. There has been resistance, including allegations that government is trying to micromanage people's lives and interfere with their freedom of choice. At this point, however, most distributors have thrown in the towel, but not USPlabs. It still maintains that DMAA is safe and can be found in geraniums that are grown under special conditions in some parts of China. But there is little question that the version in USPlabs' product is synthetic. Even if DMAA were to be found in geranium plants, extraction of the amounts used in supplements would not be possible. Furthermore, the FDA maintains that a synthetic copy of a natural constituent does not conform to DSHEA requirements. USPlabs disputes this view.

The controversy is absurd. Whether or not DMAA is sold should have nothing to do with whether it is synthetic or natural. The only question to be asked is whether it has been proven to be safe and effective. Ditto for any other supplement. As it is now, DSHEA caters not to the needs of the public, but to the whims of "Big Supplement." If a manufacturer desires to sell dietary supplements with claims of drug-like action, it should provide evidence for safety and efficacy, as is required for the sale of any other drug.

Apple Picking of Data Leaves a Bad Taste

It isn't often that I find myself in agreement with the gallant knights at the Environmental Working Group (EWG) in the U.S. who are on a quest to rid the environment of those nasty chemicals that lurk in our sunscreens, cosmetics, cleaning agents, and of course, in our food. But I'll ride along when they urge the public to eat more fruits and vegetables, even conventionally grown ones, acknowledging that the health benefits outweigh any risk posed by pesticide residues. Call me a cynic, but I think the reason that the EWG's recent press release alerting us to the "most pesticide-contaminated fruits and vegetables" led off with this bit of sound advice was to help deflect any accusations of fearmongering. But fearmongering is an apt description of EWG's list of the "Dirty Dozen" fruits and vegetables.

While the list is of very questionable scientific merit, it is undoubtedly an effective fundraiser. I have the dubious pleasure of being on the receiving end of EWG's frequent solicitations for donations: "It's important to us," they say, "to be able to continue to provide you with this cutting-edge research and easy-to-use consumer guide. Give just $10 today and we will send you EWG's exclusive shopping notepad featuring our Clean 15 and Dirty Dozen lists as a special thank you." I'm not sure that mining a U.S. Department of Agriculture database constitutes "cutting-edge research," and I'm even less sure of the usefulness of the consumer guide that is generated by cherry-picking the impressive amount of data the USDA has collected.

EWG claims that it is not out to scare the public, that it only strives to alert consumers as to which fruits and vegetables harbor the most pesticide residues and should therefore, if possible, be purchased in their organic versions. That may be the stated motive, but I suspect EWG is not averse to the donations reaped by the wide publicity the Dirty Dozen list generates. People are willing to open their wallets to support what they believe is EWG's noble effort to bring chemical criminals to justice.

Virtually every media report of EWG's recent "Dirty Dozen" press release led off with a picture of apples and a chilling headline about apples being the most pesticide-laden fruit. As a result, I fielded numerous questions along the lines of "Is it true that children should not be given apples?" "Why are apple growers allowed to profit from illness?" "How many apples can be safely eaten in a week?"

Why this focus on apples? Because they just happen to be the "dirtiest" of EWG's Dirty Dozen. Pretty convenient when it comes to garnering publicity. Apples are associated with health. After all, an apple a day is supposed to keep the doctor away. Disparaging the revered fruit in some way is almost guaranteed to generate headlines and keep EWG in the news. So how did apples end up as number one on EWG's worst offenders list? With some ingenious "apple-picking" of the data.

The United States Department of Agriculture runs a "Pesticide Data Program" that randomly tests a large variety of fruits and vegetables every year for pesticide residues. Samples are purchased across the country, washed for ten seconds, as one might do at home, and then tested for residues of the close to 200 pesticides that are registered for use. It's a stunning effort. In 2009, the last year for which data are available, over 1.5 million tests were performed on close to 11,000 samples. Of all those tests, only 1.4 percent detected a pesticide residue! I would call that pretty comforting.

Now, for the apples. In all, 744 samples were collected from across the country and tested for 194 pesticides. A total of 140,881 tests were performed, with residues being detected in only 3,717 cases. Of course, the presence of a residue does not equate to the presence of risk. The question that needs to be asked is how many of these tests found a residue in excess of the Environmental Protection Agency's carefully established maximum tolerance level? And the answer is, out of 140,881 tests . . . two!

So, given that excess residue was found in only 0.0014 percent of tests, how do you make apples the number one villain? By mining the data for that rare golden nugget of information that will dazzle the public. And it turns out there is one department in which apples really do lead the pack. Yes, 98 percent of apples have some pesticide residue! But it is not the frequency of pesticide residue detection that is important; it is the total amount of pesticides detected. EWG, however, focuses mainly on the number of times a pesticide is detected rather than calculating amounts of residue.

Here's an analogy. You're given a choice of piggy banks but are told only the number of coins each contains, not their denomination. Would it be reasonable to just choose the one with the most coins without asking any further questions? Wouldn't it make more sense to try to find out the actual types and numbers of coins? So it is with pesticide residues: the real measure of risk is determined not by how frequently residues are detected, but by how the amounts detected compare with the maximum tolerance level. And those tolerance levels are determined by finding the maximum dose that causes no observed adverse effects (NOAEL) in animals and then building in a hundred-fold safety margin for humans.

The Pesticide Data Program lists the EPA tolerance level for each pesticide and also lists the amount detected. A simple calculation reveals that, for apples, the average amount of pesticide residue detected is 2.5 percent of the EPA tolerance dose. That isn't exactly a bloodcurdling statistic, is it? But nevertheless, the "Dirty Dozen" may scare people away from eating apples. The fact is that there is absolutely no need to worry about eating them, conventional or organic. An apple a day may not keep the doctor away, but a dose of perspective about pesticide residues will keep the Environmental Working Group's fearmongering at bay.

Deer Antlers Could Have Athletes Skating on Thin Ice

Back in 1961 I had dinner with the Montreal Canadiens. Well, not exactly with them. But I did eat in the old "Texan" restaurant, across the street from the hallowed Forum, at the same time that my boyhood idols, "Boom Boom" Geoffrion, Jean Beliveau, Dickie Moore, Bill Hicke (my favorite), Jacques Plante, and the rest of that legendary team were digging into their pre-game meal. And I vividly remember what they were digging into: steak! That was standard fare for athletes at the time. The more protein, the better the prospect for butt-kicking.

Since then, the pre-game steak has given way to carbohydrate loading, but many athletes still pump protein as part of their body-building routine. Athletes, though, aren't the only ones guzzling protein. Many diet plans, such as the Atkins regimen, push protein. Students commonly snack on protein bars. And while carbohydrates and fats are often skewered by diet gurus, proteins usually get a free ride as a healthy component of the diet. But maybe proteins have been let off the hook too easily. At least, that is the opinion of Dr. Luigi Fontana of Washington University's School of Medicine. Actually, it's more than just an opinion. Dr. Fontana has carried out studies indicating that a reduction of protein intake from the North

American average of about 1.7 grams per kilogram of body weight to 0.95 grams per kilogram of body weight can have significant health benefits.

Fontana's research focuses on calorie-restricted diets, a hot topic these days because of accumulating evidence that, at least in some species of worms, insects, and rodents, such restriction has anti-aging and anticancer effects. The consensus is that these effects are due to a reduction in "insulin-like growth factor-1 (IGF-1)," a hormone also found in humans, produced by the liver in response to human growth hormone (HGH) secreted by the pituitary gland. IGF-1's main role is to transport glucose and amino acids into muscle cells to build muscle tissue. But IGF-1 is also known to promote tumor development by stimulating cell proliferation and differentiation. Furthermore, the hormone inhibits apoptosis, the programmed death that some cells, including cancer cells, undergo. So reducing the amount of circulating IGF-1 would seem to be a good thing to do. In rodents this can be done by restricting caloric intake.

Fontana knew that calorie restriction in people also lowers IGF-1 over the short term, but nobody had investigated whether long-term restriction has a similar effect. It seemed that members of the "Calorie Restriction Society," who follow a low-calorie diet while making sure that they consume adequate amounts of all essential nutrients, would constitute an interesting group for comparison with consumers of a typical Western diet. In two separate studies, the first one lasting one year, the second six, Fontana found no difference in IGF-1 levels between the groups. In other words, calorie restriction in humans has a different effect than in rodents. But when the calorie-restricted group was compared with a group of vegans who actually had a higher calorie intake, the vegans showed a significantly lower level of IGF-1. The difference, it seems, was that the vegans consumed much less protein. For humans, protein restriction, rather than total calorie restriction, may be the key to better health and longevity.

The average North American consumes about 40 percent more protein per day than the vegans in this study and has a correspondingly higher level of IGF-1. Maybe the idea that lots of protein is good for us is outdated. Maybe the reason that vegetarians and vegans are by and large healthier may not be due to what they are eating, but rather to what they are not: namely, lots of protein. It is also interesting to note that the risk of cancer increases as Asians move from a traditional low-protein diet to a high-protein Western diet.

Professional athletes, however, are more interested in short-term benefits than long-term consequences. And over the short term, IGF-1 can help build muscle. Whether the increased muscle mass leads to increased strength is open to debate, but the World Anti-Doping Agency considers that IGF-1 can build more tissue than can ever be built by training alone, and has therefore banned the substance. But the problem is that IGF-1 cannot be detected in the urine. And that is why locker room steroids are out and deer antlers are in. Yes, deer antlers! They're a source of IGF-1. Antlers are used by stags as a weapon to protect their females from untoward advances by other males during the mating season and are cast off after the season is over. This initiates the growth of new antlers, which happens at an amazing speed, thanks to IGF-1, a major player in triggering and sustaining growth.

Once the antlers are shed, they can be processed into a dietary supplement in the form of a spray that can be directed onto the thin tissue under the tongue from where it can be absorbed into the bloodstream. The purported benefits do not only include increased muscle mass, but also better heart function, an improved immune system, formation of stronger bones, and improved blood sugar levels. According to the peddlers of deer antler products, IGF-1 also stimulates the repair of damaged nerve cells and can trigger weight loss. While these claims do have some scientific basis, they have not been demonstrated in proper trials.

Although there appears to be no short-term risk, the potential long-term consequences of IGF-1 as a factor in promoting tumor growth cannot be ignored. There also have been cases where deer antler preparations have been adulterated with steroids, methyl testosterone for example, to keep customers coming back. In spite of warnings, athletes are seduced by the prospect of increased muscles, especially when marketers start tossing around remarks about how Chinese men have long used deer antler extract to enhance their manhood, not an unappealing concept for many athletes.

I never did hear any of the conversation between the players that memorable day in 1961, but there surely would have been no talk of lockouts, urine testing, or deer antlers. Drugs and labor problems were not part of the landscape. I do recall, however, that the beer flowed freely and the cigarettes glowed brightly. It didn't seem to affect performance on the ice, though. The score that night? Montreal Canadiens 8, New York Rangers 4. Ahhh . . . the good old days.

The Questionable Wizardry of Dr. Oz

When I caught a glimpse on YouTube of the omnipresent Dr. Oz donning safety glasses and getting ready to dunk a balloon into a tank of liquid nitrogen, I thought, "Great! He's going to do some real science here!" As it turned out, that was not exactly going to be the case.

Dr. Oz is an enigma. By all accounts, he's an excellent cardiac surgeon, and when it comes to explaining physiology and anatomy to the public, he does an outstanding job. But I hope he has some good orthopedic surgeon colleagues because one day he just might take a mighty spill leaping on a bandwagon driven by one of his television guests.

When Oz used to make occasional appearances on *The Oprah Winfrey Show*, he captivated with his enthusiasm and mostly sound advice about diet and exercise. And then Oprah elevated him to godly status with his own show. But you can't fill five hours a week of television by telling people to get their butt of the couch and load their plate with whole grains, fruits, and vegetables instead of burgers and fries. So Oz has to fill time with some seductive folly. And overweight people are ready to be seduced. Dr. Oz and his producers know this. They also know that at least 40 percent of their viewers are overweight and tune in hoping to hear the latest "news" about weight loss. They hang on Dr. Oz's every word and are ready to open up their wallets to try any product that seems to impress him. Unfortunately, it seems Oz is easily impressed. "Raspberry ketone" is a case in point.

Let's rewind the YouTube segment to Oz's stunning opening comment: "The number one miracle in a bottle!" You might expect to hear words like that on a late-night infomercial, but what could prompt such exuberance from a respected physician? Actually, it's more of a who. Lisa Lynn, "a personal trainer and fitness expert," just happens to sell raspberry ketone and "swears by the supplement." It melts pounds away, she declares. And Dr. Oz was ready to show us how, using a balloon to represent a fat cell in his dramatic demo.

As chemistry and physics professors commonly do, Dr. Oz proceeded to immerse the balloon in the liquid nitrogen, causing it to shrink rapidly as the air inside condensed. "This,"

Oz explained, "is what raspberry ketone does to your fat cells." Oh, really? Who says so? The personal trainer says so, claiming she believes in the product because of her "research, research, research." She buttresses her "research" with before and after pictures of a lady who supposedly lost significant weight using raspberry ketone. She is indeed plump in the first picture and a lot slimmer in the second, which shows her in a gym with weights in her hand. So was it the raspberry ketone or the exercise? I know which way I would bet.

Oz responds by saying he has become a big fan, and kicks in with his own "research" about how raspberry ketone works. He posits that it has to do with adiponectin, a hormone that "tricks the body into thinking it is thin." Adiponectin is indeed a hormone, meaning that it acts in the body as a chemical messenger. It is secreted by fatty tissues and plays a role in fat metabolism and blood glucose regulation. Interestingly, in humans, lower levels of adiponectin correlate with more body fat, so raising levels would be a reasonable avenue to explore. That has not been done.

What has been done is a laboratory experiment in which raspberry ketone prompted fat cells to release some fat while secreting adiponectin. This in no way proves that adiponectin "tricks the body into thinking it is thin." But the petri dish study was enough for inventive marketers to introduce Adipolic Fairy, a beer yeast extract containing ergosterol, a compound that supposedly increases adiponectin secretion. "Adipolic Fairy Tale" would be a more appropriate name.

A reasonable question to ask here is how the idea of investigating raspberry ketone's effect on human fat cells came about at all. That experiment was prompted by an earlier study in mice that were fed a high-fat diet. When raspberry ketone was administered at the dose of 1 percent of food intake, it prevented weight gain. For a human, this translates to several grams of raspberry ketone a day. Of course, a mouse is not a small human, so there really is no way to extrapolate a single rodent experiment to what might happen with overweight people.

There's another pertinent question: why did anyone think of feeding raspberry ketone to mice in the first place? It's a good bet that it can be traced back to the ban of ephedrine as a weight-loss supplement because of cardiac complications. A desperate supplement industry searched for a replacement and turned to synephrine, a compound that had a molecular structure similar to ephedrine and, because it was found in a citrus fruit known as "bitter orange," could be marketed with the magical term "natural."

Although, like ephedrine, synephrine does affect metabolism, it also comes with a cloud of potential side effects hanging over its head. Some astute researchers then noted the similarity of the molecular structure of raspberry ketone to both synephrine and ephedrine and thought the compound might be worth investigating in terms of weight loss. There was no safety issue here since raspberry ketone, which is partly responsible for the smell of raspberries, was already approved as a food additive. Furthermore, the wide publicity given to low-carbohydrate diets in which weight loss was linked to the body's production of ketones augured well for potential sales.

So the stage was set for a new "natural fat-burner" despite the fact that there wasn't even a single study supporting its benefits in humans. And just how "natural" is this supplement? Lynn seems to think it comes from red raspberries. And Oz pipes in with his comment that, whenever possible, he chooses "natural," feeding into the myth that natural is always superior to synthetic.

Raspberry ketone, or 4-para-hydroxyphenyl-2-butanone,

as it should properly be called, could in theory be extracted from raspberries, but this would be impractical and expensive. Laboratory synthesis, however, is easy. Of course, whether the compound is made in the lab or extracted from raspberries is irrelevant, but the implication was that people should have more confidence in the product's safety because it was "natural."

Having swallowed Oz's hype, it was the prospect of swallowing those miraculous raspberry ketone–filled capsules that sent the multitudes scurrying to health food stores, often to find the shelves bare. Such is the power of the new Wizard of Oz. With great power comes great responsibility. Dr. Oz, whom I have no doubt wants to promote a healthy lifestyle among his viewers, should be mindful of that.

Instead, he regales his audience with "miracles." These appear with astonishing frequency on *The Dr. Oz Show*. Raspberry ketone is just one example. As one miracle fades into obscurity another one quickly takes its place. Granted, Dr. Oz, or more likely his producers, do not pull miracles out of an empty hat. They generally manage to toss in a smattering of stunted facts that they then nurture into some pretty tall tales. Like the ones about chlorogenic acid and Garcinia cambogia causing effortless weight loss. The former piqued the public's interest when the great Oz introduced green coffee bean extract as the next diet sensation. Actually, chlorogenic acid is not a single compound, but rather a family of closely related compounds found in green plants, which, perhaps surprisingly, contain no chlorine atoms. The name derives from the Greek "*chloro*" for pale green and "*genic*" means "give rise to." (The element chlorine is a pale green gas, hence its name.)

An "unprecedented" breakthrough, Dr. Oz curiously announced, apparently having forgotten all about his previous weight-control miracles. This time the "staggering" results originate from a study of green coffee bean extract by Dr. Joe Vinson, a respected chemist at the University of Scranton who has a long-standing interest in antioxidants such as chlorogenic acid. Aware of the fact that chlorogenic acid had been shown to influence glucose and fat metabolism in mice, Vinson speculated that it might have some effect on humans as well. Since chlorogenic acid content is reduced by roasting, a green bean extract was chosen for the study.

In cooperation with colleagues in India who had access to volunteers, Dr. Vinson designed a trial whereby overweight subjects were given, in random order for periods of six weeks each, either a daily dose of 1,050 milligrams of green coffee bean extract, a lower dosage of 700 milligrams, or a placebo. Between each six-week phase there was a two-week "washout" period during which the participants took no supplements. There was no dietary intervention; the average daily caloric intake was about 2,400. Participants burned roughly 400 calories a day with exercise. On average, there was a loss of about a third of a kilogram per week. Interesting, but hardly "staggering." And there are caveats galore.

The study involved only eight men and eight women, which amounts to a statistically weak sample. Diet was self-reported, a notoriously unreliable method. The subjects were not really blinded since the high-dose regimen involved three pills, and the lower-dose regimen only two. A perusal of the results also shows some curious features. For example, in the group that took the placebo for the first six weeks, there was an eight-kilogram weight loss during the placebo and washout phase, but almost no further loss during the high-dose and low-dose phases. By the time that critics reacted to Oz's glowing account, though, overweight people were already panting their way to the health food store to pick up some green coffee bean extract that might or might not contain the amount of chlorogenic acid declared on the label. As for Dr. Oz, he had already moved on to his next "revolutionary" product, Garcinia cambogia, unabashedly describing it as the "Holy Grail" of weight loss.

We were actually treated to the Grail in action. Sort of. Dr. Oz, with guest Dr. Julie Chen, performed a demonstration using a plastic contraption with a balloon inside that was supposed to represent the liver. A white liquid, supposedly a sugar solution, was poured in, causing the balloon, representing a fat cell, to swell. Then a valve was closed, and as more liquid was introduced, it went into a different chamber, marked "energy." The message was that the valve represents Garcinia extract, which prevents the buildup of fat in fat cells. While playing with balloons and a plastic liver may make for entertaining television, it makes for pretty skimpy science.

Contrary to Dr. Oz's introduction that "you are hearing it here first," there is nothing new about Garcinia. There's no breakthrough, no fresh research, no "revolutionary" discovery. In the weight-control field, Garcinia cambogia is old hat. Extracts of the rind of this small pumpkin-shaped Asian fruit have long been used in "natural weight loss supplements" Why? Because, in theory, they could have an effect.

The rind of the fruit, sometimes called a tamarind, is rich in hydroxycitric acid (HCA), a substance with biological activity that can be related to weight loss. Laboratory experiments indicate that HCA can interfere with an enzyme that plays a role in converting excess sugar into fat, as well as with enzymes that break down complex carbohydrates to simple sugars that are readily absorbed. Furthermore, there are suggestions that Garcinia extract stimulates serotonin release, which can lead to appetite suppression.

Laboratory results that point toward possible weight loss don't mean much until they are confirmed by proper human trials. And there have been some. Fifteen years ago, a randomized trial involving 135 subjects who took either a placebo or a Garcinia extract equivalent to 1,500 milligrams of HCA a day for three months showed no difference in weight loss between the groups. A more recent trial involving eighty-six overweight people taking either two grams of extract or placebo for ten weeks echoed those results. In between these two major studies there were several others, some of which did show a weight loss of about one kilogram over a couple of months, but these either had few subjects or lacked a control group. Basically, it is clear that if there is any weight loss attributed to Garcinia cambogia, it is virtually insignificant. But there may be something else attributed to the supplement, namely kidney problems. Although incidence is rare, even one case is too many when the chance of a benefit is so small. So Garcinia cambogia, like green coffee bean extract, can hardly be called a miracle. But it seems Dr. Oz puts his facts on a diet when it comes to fattening up his television ratings.

Breatharians and Nutritarians

Breatharians and nutritarians. Never heard of them? One represents the extent of human folly, while the other is a scientifically legitimate attempt to improve health. A breatharian is a person who, under the proper conditions, can live without food. Who says so? Wiley Brooks, who just happens to be the founder of the Breatharian Institute of America. "If food is so good for you, how come the body keeps trying to get rid of it?" asks this mental wizard. "Eating is an acquired habit," Brooks continues, "all of the constituents we need can be taken from the air we breathe."

Of course one must ascend to a certain spiritual level before one can forego food, and Mr. Brooks is happy to show us the way. But a consultation with this ascended master, who claims to have not eaten for over thirty years, doesn't come cheap. The minimal fee is $10,000! Just think, though, of all the money you'll be saving by not having to buy food for the rest of your life. Brooks apparently has lots of experience in advising others to attain "incredible love, peace, and joy" through living on air. You see, he has had past lives as Adam, Zeus, Jesus, John the Baptist, Joseph Smith, and curiously, William Mulholland, an engineer who designed Los Angeles' aqueduct system. Quite a puzzle even for believers in reincarnation, given that Jesus and John the Baptist were contemporaries.

You would think that Brooks is a unique looney. He's not. Ellen Greve, who has taken on the name of Jasmuheen, heads the CIA. No, not that CIA! This "Cosmic Internet Academy" is in Australia and "offers some unusual solutions to world hunger and health issues." Unusual is right. The solution to world hunger is "pranic nourishment." "Prana" is the universal life force that can provide all. It does seem to provide nicely for Greve, who charges $2,000 for her enlightening seminars about eliminating food except for tea and an occasional bit of chocolate or ice cream when she needs a "taste orgasm."

When the Australian version of *60 Minutes* challenged her to put her breath where her mouth is, that is demonstrate that she can live on an intake of "cosmic particles," she failed miserably. A physician ordered the test stopped because this sage, who had authored a book about a twenty-one-day program that allows the body to stop aging and attain immortality by living solely on light, was on the verge of proving her mortality as her kidneys began to shut down. Her rationale? The polluted city air was void of nutrients.

Jasmuheen lived to see another day after being rehydrated, but the same cannot be said for four unfortunate souls whose deaths from dehydration have been linked to their having followed her zany publications. Jasmuheen explains that such tragedies can occur if you haven't found the light that will nourish you.

Prahlad Jani, an Indian guru, apparently has found the divine light. He claims to have lived without food or water since 1940. Jani spends his time living in a cave but has twice emerged to be tested by physicians who claim he did not eat, drink, urinate, or defecate during a two-week observation period. Their account has not been published and has been dismissed by experts who claim that the guru was in fact not observed at all times. But let's not waste any more breath on the nonsense of breatharianism. While we could all eat less, the intake of food and water for humans is not optional. But, of course, magical thinking is. So is fraud.

You won't find any breatharians at the Pyramid Bistro in Aspen, Colorado, but you will find some nutritarians. They're a breath of fresh air when compared with breatharians. Nutritarianism may also sound like some strange cult, but it isn't. There is, however, some "worship" involved, that of "nutrient density." The term "nutritarian" was coined by Dr. Joel Fuhrman, a family physician who believes that many diseases can be prevented, or even cured, by eating nutrient-dense foods. Fuhrman recommends a diet based on the Aggregate Nutrient Density Index, or ANDI. The index compares the nutrients a food contains to its calorie content, and assumes that the higher this ratio, the "healthier" a food is.

Just what do we mean by nutrients? Generally, food components can be divided into macronutrients and micronutrients.

Fats, proteins, and carbohydrates provide the building blocks for our body and also serve as our source of energy. Vitamins, minerals, and numerous other molecules that are present in smaller amounts, but have biological activity, constitute the micronutrients. Antioxidants such as beta-carotene in carrots, lycopene in tomatoes, or anthocyanins in blueberries are prime examples.

Among green vegetables, kale, watercress, and bok choy top the ANDI list, while strawberries and blackberries lead the fruit pack. Beans rank high, as do sunflower, sesame, and flaxseeds. So do whole-grain oats. The nutritarian diet is not totally vegetarian, but about 90 percent of the content comes from nutrient-rich foods such as beans, seeds, nuts, mushrooms, and fruits and vegetables, especially onions, berries, and leafy greens. Compare this with the standard North American diet, in which only 5 percent of calories derive from these foods.

Fuhrman's official blog is entitled "Disease Proof." A little over the top, I would say. There are also excessively optimistic statements about preventing heart disease and cancer, and suggestions that "you don't have to live the rest of your life in pain or on medication." Interestingly, while the dietary regimen is supposed to maximize micronutrient intake, Fuhrman sells dietary supplements. Still, living by the nutritarian credo is certainly preferable to the usual North American diet, and judging by the reviews I've seen about the fare at the Pyramid Bistro, it can be delicious.

I'm keen to try the flaxseed-spelt gnocchi with tomatoes, English peas, snap peas, mustard greens, sunflower seeds, and aged balsamic, or the lemongrass tofu forbidden rice, with ginger and steamed bok choy in a spicy carrot emulsion. Forbidden rice isn't actually illegal. It is so-called because nobles in ancient China thought it so valuable that they commandeered all that could be grown for themselves and forbade its consumption by commoners. Maybe a research trip to Aspen is in order. I hear even breatharians go there. Seems the unpolluted air is especially nutritious. I would really like to meet one of these science-defying wonders of nature. But I'm not holding my breath.

Pink Slime — Jamie Oliver Chooses the Wrong Bone to Pick

Jamie Oliver doesn't like "pink slime." He doesn't want any of it in his hamburger. In fact, the famed British chef was so disgusted that McDonald's in the U.S. was using this "beef filler" that he orchestrated a campaign to get rid of it. So what is "pink slime," as Jamie calls it, and what horrors does it hold?

Once a cow has been butchered and disassembled into the various cuts, some fatty trimmings always remain. Traditionally, these have been used for pet food, but in 1991, an American company, Beef Products Inc., developed a process to convert the trimmings into what it calls "Boneless Lean Beef." The fatty portions are separated by spinning in a sophisticated centrifuge, leaving behind the muscle tissue, which is then ground into a slurry that is roughly 94 percent lean beef. This "pink slime" is then frozen into chips or blocks, ready to be incorporated into hamburger or into processed luncheon meats.

Because meat trimmings are particularly susceptible to bacterial contamination, the company introduced a novel method to control the risk. Salmonella and E. coli bacteria, having evolved in an acidic environment, cannot survive under basic, or alkaline, conditions. Ammonia gas can introduce such conditions as it dissolves in water to form ammonium hydroxide, a base. An equilibrium is then established between the dissolved ammonia and the ammonium hydroxide, with very little ammonium hydroxide actually present at any given time. But when it is used up, as in the reaction to destroy bacteria, it is replenished as more of the dissolved ammonia reacts with water.

Ammonia gas is used to treat the beef slurry as it passes through specially designed stainless-steel pipes. Some of the ammonia dissolves in the meat's moisture and maintains the alkaline conditions needed to control bacteria. Neither the dissolved ammonia nor the ammonium hydroxide it forms presents a health concern. Ammonia is a product of protein metabolism and therefore routinely forms in the human body. It ends up being converted into urea, which is then excreted in the urine.

Since alkaline solutions are very effective at breaking down greasy materials, dissolved ammonia is widely used in cleaning agents. Many window-cleaning products feature ammonia as their basic ingredient. And when Jamie Oliver made his version of "pink slime" on TV, cleaning agents were prominently featured. In a cleverly crafted "made for TV" piece, Jamie opened up a padlocked cabinet, obviously intended to emphasize the danger, and removed a bottle of ammonia that prominently featured the skull-and-crossbones symbol. I've never seen such a bottle. He then proceeded to place some meat trimmings in a washing machine that played the role of a centrifuge, and dumped in the ammonia cleaner liberally while admitting he has no idea how much to use. The impression given was that the meat is washed in an ammonia solution, which is not at all the case. Jamie's ugly mash solicited plenty of "yucks" and "ewwws" from the onlookers, along with snide comments about the food industry.

Let's get real here. Whether or not ammonia is found in cleaning agents has nothing to do with whether it is safe or effective as an antibacterial agent in meat. We don't worry about salt in our food because it is a substance that is also spread on streets to melt ice. We worry about it because studies have shown that at some doses, it can cause problems for some people. Ammonia in meat production has been well studied and it is known that the amount added to "boneless lean beef" does not leave a residue we need to be concerned about. There is, however, some concern that the amounts used may not offer as much protection against bacteria as claimed. And if more is used, the meat develops an "off" taste.

Revulsion of hamburger because an "industrial cleaning agent" is used in processing is not warranted. Neither should the fact that meat trimmings, instead of being used for dog food, are centrifuged into a slurry that ends up in some burgers provoke disgust. When the issue of "pink slime" arises, the appropriate question is whether hamburger made with it differs significantly in terms of nutrition and safety from hamburger made without it.

Perhaps surprisingly, this is not a difficult question to answer. McDonald's in the U.S. has used "pink slime," McDonald's in Canada has not, and the nutritional profile of the company's products is readily available. The fat and protein content in the American and Canadian burgers is the same. So it seems the use of the highly processed mash doesn't make a difference. And why should it? The processing actually removes fat from the trimmings. What is then added to the burger is no more fatty than the rest of the meat that is used.

Jamie Oliver professes not to be against eating hamburgers; he says he just wants to know where the ingredients come from. And "pink slime," he says, is not fit for human consumption. Well, how does one determine if a food is fit or not? By its microbiological safety and its chemical composition, not by its origin! That's more a matter of social and cultural views. Most North Americans don't find haggis or snake stew very appetizing, but don't mind gnawing on the ribs of a cow.

Oliver's crusade for better nutrition is admirable. But imploring people to revolt against the use of ammonia-treated beef slurry is a misguided attempt to improve eating habits through scare tactics. Without a doubt, though, such tactics can achieve results. McDonald's has stopped using pink slime, although the company says the decision had nothing to do with Jamie Oliver's crusade. Right. Is hamburger made without pink slime in any way healthier? No. Hamburger is a fatty, salty food that should be consumed in limited amounts whether it is made with pink slime or not.

Actually, these days, with a push to reuse and recycle, one could argue that methods to convert unusable fatty meat remnants into usable lean beef should be pursued instead of reviled. Basically, more meat is produced with less feed. Given that there's no safety or nutritional issue, why is that a problem? Jamie claims that such processing shows "no respect for food or for people." Why is a more efficient use of butchered animal disrespectful? In any case, killing animals in order to eat them isn't exactly respectful in the first place, is it?

Quackery Can Tarnish Silver's Medical Luster

A child born into a wealthy family is often said to have been "born with a silver spoon in his mouth." The silver represents wealth, but it may even have a connection to health. That's thanks to the oligodynamic effect, discovered in 1893 by the Swiss botanist Karl Wilhelm von Nageli. This refers to the toxic effect of metal ions on living organisms such as bacteria, algae, and fungi. Copper, lead, zinc, gold, aluminum, and mercury can also furnish ions that produce an antimicrobial effect, but the potential benefits have to be weighed against the metals' toxicity. Mercury at one time was commonly used to treat syphilis, giving rise to the expression, "a night with Venus, a lifetime with mercury." However, because of mercury's toxicity, that lifetime was often quite short. Copper and silver, on the other hand, have low toxicity by comparison and can be effectively used to battle bacteria.

As early as the ancient dynasties of Egypt, silver coins were placed in the drinking vessels of the nobility to protect them from harm. Of course, this was not the result of any scientific investigation; the practice probably originated from some superstitious belief about the magical properties of precious metals. Over the years, it became apparent that the silver coins really did have an effect: they kept water from becoming slimy. Storage of water in silver vessels was an obvious extension of this observation, offering the well-to-do some protection from water-borne diseases that were common before the introduction of chlorination. As recently as the twentieth century, the Maharaja Sawai Madho Singh II journeyed to England with two sterling-silver vessels filled with holy water from the Ganges River. The vessels are acclaimed to be the largest silver containers ever produced.

You do not have to be as rich as a maharaja to experience the oligodynamic effect of silver. The metal can be incorporated into urinary catheters and endotracheal breathing tubes to reduce infections, and fabrics can be formulated with small amounts of silver to control the bacteria responsible for churning out odorous compounds when they feast on sweat. The antimicrobial effect is actually due to silver ions — in other words, silver atoms that have lost an electron. These ions are produced whenever silver atoms at the surface of the metal react with either oxygen or hydrogen sulphide, the "rotten egg" compound that is always present in the air and water in trace amounts. Indeed, the "tarnish" on silver is silver sulphide, the product of the reaction between silver and hydrogen sulphide.

Silver ions inactivate enzymes that are essential for bacterial life. That's why bacteria are killed when contaminated water is stored in a silver container. But in this case, the extent of disinfection is unreliable because the concentration of silver ions in the water cannot be controlled. The purity of the silver, the size of the container, and whether the water is shaken are each important determinants of the concentration of silver ions. However, techniques have been worked out to produce just the right concentration of ions by immersing a pair of silver electrodes connected to a direct current into water that needs to be purified. This was the method used to produce drinking water aboard the Apollo space flights and is used in hospital plumbing systems to deactivate Legionella bacteria. Copper is often alloyed with silver in the electrodes to take advantage of its oligodynamic effect as well. Swimming pool disinfection systems using copper-silver ionization that allow for reduced use of chlorine are also available.

If silver ions produced on the surface of the metal are the active disinfecting agent, it stands to reason that the surface area of the silver would play an important role in the effectiveness of the treatment. And it does. Particles of silver that are less than a billionth of a meter in size, commonly referred to as nanoparticles, have been shown to be especially effective at killing bacteria. A recent study by Dr. Derek Gray at McGill University showed that passing contaminated water through absorbent blotting paper treated with silver nanoparticles resulted in inactivation of bacteria. This has the earmarks of a landmark discovery because the consumption of contaminated water in the developing world is a major health crisis. Nanosilver-impregnated paper is easy to produce and easy to use.

Unfortunately, quackery often rides along the coattails of science. And so it is in this case. Numerous websites promote the use of "colloidal silver" as a cure for cancer, diabetes, HIV infection, and herpes. Colloids refer to a system in which finely divided particles are dispersed within a continuous medium without settling out. In the case of colloidal silver, these particles can be elemental silver or particles of silver compounds. Indeed, they may well have an antibacterial effect in a petri dish, but that is a long way from having an antibacterial effect when taken internally. No scientific evidence supports the benefit of ingesting any form of colloidal silver. Making health claims on its behalf is illegal, but colloidal silver can be sold as a dietary supplement. That is a curiosity because humans have no dietary requirement for silver, and there is no such thing as silver deficiency.

But there certainly is such a thing as silver excess. The metal can deposit in the skin as well as in internal organs, and the result is a condition known as argyria. Its hallmark is gray-blue skin, a condition that is irreversible. One of the most famous cases was that of the Blue Man, who was a featured attraction in the Barnum and Bailey Circus in the early years of the twentieth century. He had apparently tried to cure his syphilis by ingesting silver nitrate, but succeeded only in making himself blue.

More recently, Stan Jones, an American Libertarian who twice ran unsuccessfully for the Senate as well as for Governor of Montana, did succeed in becoming blue. On the cusp of the year 2000, he was worried that computers would stop functioning and that this would somehow make antibiotics unavailable. He decided to take preventive action and started to take a colloidal silver preparation that he made himself by passing an electric current through a solution equipped with silver electrodes. Unfortunately, he didn't know what he was doing, he used too high a voltage, and his solution contained a great deal of silver. He turned blue. But he is not singing the blues. He maintains that he is healthy, and still dopes himself with colloidal silver. Pretty dopey, actually.

Salt Therapy

In Canada, we're accustomed to spreading salt on our roads to melt ice. But according to the promoters of halotherapy, it may be a good idea to salt your lungs as well. They're not talking about snorting salt. Not directly, anyway. Halotherapy is all about breathing in the salty air while relaxing in a "salt room," where the floor and walls are lined with crystals of sodium chloride. Why? Supposedly to ease breathing problems, allergies, and ear infections. Salt rooms are cropping up in major cities, including Toronto, featuring a "microclimate" that is said to resemble that found in salt mines, which, at least according to anecdotal evidence, has a beneficial effect on health. Maybe the ancient Romans who punished prisoners by sending them to the salt mines were actually doing them a favor!

Why did the Romans have salt mines? Two reasons: first, salt was essential for preserving food and was also an effective weapon. "Salting the earth" destroyed the enemy's agricultural fields and consequently, its food supply. Salt was so important that Roman soldiers were given extra money to purchase it. This was referred to as a "salarium," from which our word "salary" derives. Soldiers who performed their duties particularly well were said to be worth their salt!

Although today we mostly associate excessive salt intake with high blood pressure, the use of salt as a therapeutic agent has a long history. Indeed, 2,500 years ago, Hippocrates suggested immersion in salt water for various ailments. Numerous published articles refer to the therapeutic effects of immersion in the Dead Sea, as well as other saltwater bodies, particularly for chronic skin conditions. Because this therapy also involves prolonged exposure to sunlight, it is not clear to what extent the benefits are due to the water, the various minerals, the high salt content, the sun exposure, or to a combination of factors. Maybe even to breathing the salty air.

Back in the nineteenth century, Polish physician Feliks Boczkowski noted that workers in salt mines seemed to suffer fewer lung problems. A similar observation was made during World War II in Germany by Dr. Karl Hermann Spannage, who claimed that his patients who hid in salt caves to escape Allied bombing saw an improvement in their health. Word spread and physicians began to recommend that patients with breathing problems spend time in caves where the salty air, free of pollen and pollutants, was said to do them a world of good. "Speleotherapy" soon became popular in Eastern Europe, particularly in the Ukraine.

Popularity, however, does not equate to efficacy. Claims of improvement by patients who have spent time in Ukrainian salt caves do not constitute scientific evidence. Determining whether a medical intervention actually works requires carefully controlled studies to eliminate the placebo effect. Halotherapy does have some scientific plausibility, given that salt is known to have antibacterial properties. After all, that is how it acts as a food preservative. But whether the amount inhaled in salty air is enough to have a physiological effect is a different story.

The only study that can even remotely be called scientific was published back in 1995 by a Russian researcher. The placebo group consisted of patients who just sat in an ordinary room listening to music as they watched some sort of slide presentation. That's a very different setting from a salt cave. Both the experimental and control groups should have been asked to spend time in caves after being told that there was something special about the air they would be inhaling, with only the experimental group actually being exposed to salt. There was yet another problem with the study: most of the patients were also on some kind of medication, making it difficult to determine if any improvement was due to the salty air.

In any case, the Russian study did not yield impressive results. Although there was an overall trend toward improvement in the salt cave, some patients actually got worse. Not really surprising. Inhaling microscopic particles of anything can impair lung function. Aside from this poorly designed study, there isn't much published on salt therapy. A paper published in the *New England Journal of Medicine* in 2006 did describe research that was apparently stimulated by surfers in Australia who had noted an improvement in cystic fibrosis–associated breathing problems when they inhaled sea spray. The researchers found that inhaling a salt solution improved lung function in cystic fibrosis patients. The same year, a publication in the *European Respiratory Journal* suggested that inhaling aerosolized salt temporarily improved smoking-related coughing and mucus production. Of course, giving up smoking will improve these symptoms to a far greater extent.

As far as salt rooms go, evidence is essentially anecdotal. That doesn't mean it should be dismissed. The pursuit of scientific evidence often begins with an anecdotal observation. But it must progress beyond nebulous claims about how "low concentrations of salt are delivered to the lungs where the salt then dissolves phlegm and kills microorganisms that cause infections." This may indeed be so, but there is no documented evidence to back up the claim. Still, when hay-fever sufferers and asthma patients report that they feel better after their salt room sessions for which they forked out some sixty dollars per hour (maybe they would feel even better if the price were raised), scientists should take note and mount a proper controlled trial. After all, salty air is less invasive than steroid therapy, and inhaling it while sitting in a comfortable chair and listening to pleasant music may actually be therapeutic, even if the salt plays no part.

Red flags do, however, go up when claims are made about treating conditions such as ear infections and chronic cough. There is no evidence that salt rooms can cure these, and a belief that they might could lead a patient to put off more effective treatment. Chronic cough can be a symptom of a serious condition that requires more than salt, New Age music, and a comfy couch. I'd also welcome some evidence supporting the claim that salt rooms help reduce snoring. Earplugs might be a better investment. And as far as claims about enhanced sports performance go . . . maybe, if they install a treadmill in the salt room.

The Funny Business of Selling Water

When in New York, you might want to drop in at the Molecule Café in Greenwich Village. Just don't look for any coffee with your cake. In fact, don't even look for any cake. The only item the café serves is water. And at $2.50 a glass, it's no bargain. You can toss in another two bucks for a dash of an "infusion." But the large dose of unsubstantiated hype that comes along with your purchase is a bargain: it's free. The "infusions" include a cacophony of vitamins, minerals, herbal extracts, and neurotransmitter precursors that are supposed to improve your appearance, immune system, and mental function. The latter claim is highly questionable, seeing that customers keep going back to spend an outrageous amount of money for something that is available from the tap for free.

Of course, the sales pitch delivered at the Molecule Café suggests that tap water harbors "toxic" chemicals that may precipitate an early date with the undertaker. The usual suspects accused of villainy include chlorine, fluoride, pesticides, trihalomethanes, and solvents such as isophorone, used in printing inks. Yes, these chemicals may be present in water, and it is true that they can all cause problems upon significant exposure, but maximum levels allowed in water have been established through extensive scientific investigation. Municipal water treatment facilities monitor for a large array of chemicals on a daily basis, with New York City carrying out more than half a million tests a year. Traces of isophorone, for example, may indeed show up. This, however, has no relevance because the highest level of this compound ever found was 500 times less than the safety limit of 5 milligrams per liter, which already has a thousandfold safety factor built in to correct for extrapolation from animals to humans. Indeed, you are more likely to encounter isophorone in cranberries, where it occurs naturally.

Adam Ruhf, who boasts a background in music and "activism," is the brains behind Molecule Café. He maintains that he doesn't want any "toxins" in his water, claiming that his recovery from a serious car accident was aided by the healing powers of pure water. And his establishment does provide pure water. Ruhf has installed an elaborate purification system to basically kill an ant with a jackhammer.

There's no doubt that the combination of reverse osmosis, activated carbon, kinetic degradation fluxion, 0.0002 micron filter membranes, and ozone treatment adds up to very high-quality water. Of course, the question is whether this really matters in light of all the chemicals to which we are exposed in our daily lives. Take a whiff of gasoline vapor when filling up your tank, bite into a piece of burnt toast, sip beer, taste some brown rice, or sniff bleach, and you'll be exposing yourself to benzene, acrylamide, ethyl carbamate, arsenic, and hypochlorous acid, all of which can be described as "toxic." So, dropping in to the Molecule Café for a glass of ultra-pure water is not the answer to toxin exposure. It may not even do much for the taste buds. In blind trials, volunteers were unable to distinguish Molecule water from New York City tap water, which always gets high marks from tasters.

To the establishment's credit, there is no bottled water in sight. You can bring your own containers to be filled or buy refillable glass bottles, but there are no stacks of bottled waters to cart away. That's commendable because bottled water is a superfluous product. In North America, we consume some 40 billion liters of such water a year, with roughly 1,500 caps being twisted off every second! And while we talk about the benefits of eating locally, and raise concerns about importing produce from Chile or China, the far more significant impact of transporting water from Europe or, from of all places, Fiji, seems to float under the environmental radar. Producers, stores, and restaurants don't mind. The profit margin on bottled water is huge, with many versions selling for more than gasoline with a fraction of the production costs.

And then there are the bottles themselves. The plastic that is used is polyethylene terephthalate (PET), number one on your recycling logo. There are some real issues here concerning production and disposal. Historically, the raw materials, namely ethylene glycol and terephthalic acid, are derived from fossil fuels, a waste of a valuable resource. And while PET can be recycled, the overall recycling rate is low, with the majority of bottles being discarded improperly.

Marketers of water and soft drinks have reacted to accusations of unnecessarily depleting fossil fuels by engaging in extensive research to produce the needed plastic from renewable resources, namely plants. Ethylene glycol can be made from ethanol, which in turn can be produced by fermentation of sugars, but only 30 percent of the weight of the much-ballyhooed "plant bottle" comes from ethylene glycol. Terephthalic acid still has to be sourced from petroleum, although it may eventually be produced from plant sources as well.

One possibility is to use specific microbes to ferment sugar into isobutanol, which can then through a series of reactions be converted to paraxylene, the essential precursor for terephthalic acid. There are also catalytic methods for making paraxylene from sugar, but no matter what, all methods involve a great deal of chemical processing. Sourcing the raw material from plants makes for good advertising copy, but is hardly the solution to the problem of flooding the world with plastic bottles that contain beverages we could easily do without. The industry talks about decreasing the amount of plastic in each bottle and about efforts to promote recycling, but remember that a bottle that isn't made doesn't use up resources and requires no recycling.

Anyone concerned with the quality of tap water can easily avail themselves of a variety of filters that will produce water comparable to the bottled variety at a fraction of the cost. Whether this has an impact on health is debatable, but removal of traces of chlorine can improve the taste. Still, I suspect this filtered water won't radiate the same aura as the $2.50 glass at the Molecule Café, especially when boosted with an infusion of "Fountain of Youth" and blessed by a Tibetan monk, as is in the works.

Buckyballs Roll into the Pit of Folly

The most memorable remaining landmark from Montreal's fabulous Expo 67 is the giant geodesic dome designed by architect, engineer, and futurist Buckminster Fuller (1895–1983) for the U.S. pavilion. The magnificent dome, 62 meters high, now houses an environmental museum known as the "Biosphere." Fuller, who dreamed of energy efficient homes, recycling, and global sustainability long before these ideas became fashionable, would be pleased. But the famous inventor, writer, and designer surely never dreamed that his name would be immortalized in numerous chemistry journals, lectures, and textbooks, let alone as the name of a substance. Buckminsterfullerene is a fascinating substance, important enough to have its discovery recognized with the 1996 Nobel Prize in chemistry, awarded to Drs. Richard Smalley, Harry Kroto, and Robert Curl.

It was back in 1985 that the three researchers made a curious discovery when using a special laser to vaporize graphite. The intense heat of the laser caused the substance to decompose into a number of products, one of which corresponded to a species having sixty carbon atoms and nothing else. How sixty carbon atoms, each of which can form four bonds, could be joined into a stable structure was a real puzzle.

While there's no longer any doubt about the answer to this puzzle, there is controversy about just how the puzzle was solved. Kroto's and Smalley's recollections of the brainstorming that took place differ significantly. Kroto claims that it was a memory of his visit to Fuller's geodesic dome at Expo 67 that triggered the idea of the sixty carbon atoms joined together in the shape of a soccer ball. Smalley, who passed away in 2005, said he arrived at the structure by making paper cutouts of hexagons and pentagons representing carbon atoms at each corner and fitting these together into a spherical shape. There was sufficient disagreement over this to cause a personal falling-out that was later resolved. But there was no disagreement about paying homage to Buckminster Fuller by naming the novel substance "buckminsterfullerene," which the lay press affectionately shortened to "buckyball."

Almost immediately after its discovery, buckyball research got rolling in labs around the world. It turned out that not only could carbon atoms assemble into hollow spheres, they could also join to form ellipsoids and nanotubes. At first, significant amounts of these fullerenes were hard to come by, but before long chemists had discovered that buckyballs occurred naturally in soot and techniques were quickly worked out for mass production. There was no doubt that the soccer ball-shaped molecules were theoretically interesting, but of what practical use were they? There were hints of super strength, superconductivity, and even of various medical applications. Oddly, and perhaps appropriately given its name, the first commercial item to incorporate buckyball technology was the "Nanodesu" bowling ball, manufactured in Japan. The fullerene was added to the polyurethane coating of the ball to improve its "controllability."

While the strength and conductivity of fullerenes hold lots of potential, it is their medical applications that excite researchers. Buckyballs can be armed with anti-tumor antibodies and then assembled into aggregates called buckysomes that are packed with anti-cancer drugs. Instead of attacking all cells, these complexes bind only with receptors on tumor cells before releasing their load of therapeutic drugs. Buckyballs also have free-radical-scavenging activity, anti-viral effects, immune-stimulating properties, and even hair-growing potential.

In drug therapy, beneficial effects are always burdened with the possibility of toxic side effects. In order to explore the potential long-term toxicity of buckyballs, researchers at the University of Paris treated rats with periodic doses of 1.7 milligrams per kilogram of body weight until the end of their days. Not only were there no toxic effects observed, but the lifespan of the treated rats was almost double that of the controls! The supposition is that this is due to the free radical–scavenging activity of buckyballs, based on a separate experiment in which rats were treated with carbon tetrachloride, a chemical known to damage the liver by inducing the formation of free radicals. Impressively, pre-treating the animals with a buckyball solution in olive oil protected the liver against carbon tetrachloride toxicity. Interesting, but the study has not been repeated and its methodology and results have been criticized.

Leave it to the quacks to step in and hijack the science, questionable as it may be, with the promotion of a nonsensical product called "C60 Water of Life." Just drink a few spoonfuls of this wonder water every day, we are told, and it increases energy levels; reduces the risk of cancer; fights stress, depression, and chronic fatigue; protects the liver; provides effective protection against radiation, colds, and flu; heals burns and ulcers; provides long-term antihistamine and anti-inflammatory effect; prevents buildup of deposits in arteries; and even inhibits menopause. Needless to say, it reduces the side effects of chemotherapy and shortens the duration of treatments needed for multiple sclerosis, Alzheimer's, and Parkinson's disease. What more can one ask for? Hello! How about a little evidence? The inventive people who have produced the brochure hyping this miracle tell us that the curative properties have been confirmed by the Ministry of Health of the Ukraine, although no documentation is provided for that claim.

So what does C60 Water of Life contain? Water. And a vanishingly small amount — two parts per billion — of buckminsterfullerene, around which the makers of C60 Water of Life weave their tangled web of deceit. "This product is composed of highly purified water in which natural structures — spherical water clusters — are stabilized with the help of molecular carbon. The penetration of carbon molecules allows the clusters to live for an indefinite period of time. In addition, the spherical carbon molecule is surrounded by ordered layers of water, like water in the human body. Restoration of human health should start with the restoration of the internal water of your body." What we have here are a few smidgens of scientific fact that are inflated and distorted to form a giant glob of scientifically distasteful nonsense that nevertheless sounds palatable to the gullible.

Remember that the study on which the poppycock is based was a rat study, the methodology and results of which have been called into question. And the "active ingredient" was dissolved in olive oil, not suspended in water. Fullerenes are known to be biologically active only in solution, not as a suspension. Furthermore, the rats received about 10,000 times more buckminsterfullerene a day than that present in the recommended dose of Water of Life. Eventually fullerenes may prove to have real medical value, perhaps even in increasing longevity, but as far as "C60 Water of Life" goes, it deserves a quick death.

Scientists Smell a Rat in French GMO Rat Study

A French study published in 2012 that purports to show a link between the consumption of genetically modified corn and a variety of ailments, including cancer, was just the tasty morsel that critics of genetically modified foods (GMOs) hungered for. For many scientists, however, the study proved to be a source of indigestion.

Although California's Proposition 37, which would have required the labeling of foods that have any component derived from genetically modified crops, was defeated in November 2012, GMOs are still a hot-button issue. Emotions have boiled over with members of activist groups, such as the ridiculously named "Genetic Crimes Unit," screaming about genetic crimes against humanity as they don hazmat suits to block shipments of Monsanto's transgenic seeds. They are also fond of displaying a giant "fish-corn," implying that biotechnology companies are engaged in melding fish genes with corn genes. Absurd.

Mike Adams, the self-appointed "Health Ranger" who routinely floods the internet with stupefying diatribes on his "Natural News" website, goes even further. "I predict, but DO NOT CONDONE," he says, "scientists who conduct research for Monsanto being threatened, intimidated, and even physically attacked . . . an inevitable reaction to the unfathomable evil being committed by the GMO industry and its co-conspirators." Seems to me that Adams is the evil one by implanting such ideas. There are indeed some very legitimate issues to be addressed about genetic modification, but proper intellectual discourse leaves no room for such inflammatory tirades.

Mistrust and confusion are often the result of a lack of understanding of the science involved. So let's take a look at what the controversy, at least as it pertains to the French study, is all about. The researchers aimed to explore the effects of consuming corn that is genetically modified to resist Roundup, Monsanto's popular herbicide. Such Roundup-resistant corn is unharmed when sprayed with glyphosate, the active ingredient in Roundup, while weeds wilt. This is of great advantage to growers because the technology makes weed control easier and more effective, and fields require less tillage while yields and profits increase. Before the introduction of glyphosate-resistant crops, it was common to use as many as ten different herbicides, most of which had worse toxicological profiles than glyphosate.

Glyphosate was discovered by John Franz back in 1970, while he was working at Monsanto. It works by inhibiting the plant enzyme EPSPS (5-**e**nol**p**yruvyl**s**hikimate-3-**p**hosphate **s**ynthase, if you must know) which is critical for the synthesis of three essential amino acids: tryptophan, tyrosine, and phenylalanine. These in turn are needed by the plant for protein synthesis as well as for conversion into a variety of compounds such as phenolics, tannins, and lignins that are essential for plant life. If EPSPS is inactivated, the plant withers and dies.

Some microbes also rely on EPSPS for protein synthesis, and in 1983, researchers discovered that a strain of the common soil bacterium *Agrobacterium tumefaciens* is highly tolerant to glyphosate because its EPSPS is less sensitive to inhibition by this herbicide than the version found in plants. By 1986, the bacterial gene that codes for this enzyme was isolated and soon inserted into the genome of soybeans, corn, canola, alfalfa, and sugar beets, allowing fields to be sprayed with Roundup for elimination of weeds without affecting the crops. As a result, genetic modification has become the most rapidly adopted technology in the history of agriculture. But it has also unleashed a cavalcade of criticism.

There are concerns about seed companies establishing strict criteria for the use of their seeds by farmers, there are questions about weeds developing resistance, and, of course, worries about safety. While the majority of scientists familiar with the technology were satisfied that the concerns had been properly addressed, there were some who thought that regulatory agencies had jumped the gun. One of these was Gilles-Eric Séralini, lead author of the current controversial study.

Séralini has written several anti-GMO books and has published other papers that claim to show adverse effects attributed to GMOs. He is a vocal anti-GMO activist and has already been chastised by the European Food Safety Association (EFSA) for improper analysis of data. His present study involved feeding various combinations of genetically modified corn and glyphosate to rats over their lifetime and concluded that the experimental rats had a shorter life expectancy, developed more tumors, and had more liver and kidney problems than the control group. There were horrific pictures of rats with giant tumors that were quickly snapped up by a media not adverse to sensationalism.

The response from the scientific community was immediate and harsh. The control group was way too small, there was no disclosure of control rats with tumors, data were improperly interpreted, there was no dose-response relationship, and the strain of rat used was genetically susceptible to tumors. Particularly bothersome was the fact that the research group refused to provide advance copies of their work to reporters unless they signed agreements not to consult other experts. This flies in the face of proper scientific practice. Furthermore, Séralini has now stated that he will not allow scientists from EFSA to verify his results because they are the ones who approved GMOs in the first place and therefore cannot be trusted. That sort of behavior, to be kind to the man, is bizarre.

The true crux of the matter is that this study has virtually no relevance to people because the diet the rats were fed is not even remotely reflective of the human consumption of foods that have components derived from genetically modified corn. The media randomly bandies about the statement that most of the food we eat contains genetically modified ingredients. Technically, that is true if you consider, for example, high-fructose corn syrup (HFCS) derived from GM corn to be a genetically modified ingredient. The fact is that there is no vestige of genetic modification in this product. It is indistinguishable from any other HFCS. Contrary to popular belief, there are no genetically modified strawberries, tomatoes, potatoes, wheat, rice, or fruits on the market, with the exception of Hawaiian papaya, which has been engineered to protect it against a fungus, thereby saving a whole industry.

Although GM sweet corn is grown in a few places, by far the majority of GM corn goes into animal feed. Our consumption of GM ingredients is limited to some food additives and oils that are derived from GM corn, soy, or canola. This has little relation to feeding GM corn to rats as the major component of their diet. Furthermore, millions and millions of cattle and poultry have now been raised on GM corn over many generations without any health effects being noted in them or their consumers.

What we need in the GMO controversy is reasoned argument, not scandalous headlines. "Study: GMOs may shorten your life" shrieks a report on Séralini's paper by Rodale Press. The study shows nothing of the kind. What it does show is the readiness of some GMO opponents to jump on a questionable study to promote their fearmongering agenda.

WHITE

It's in the Can!

It may not be quite on par with the Manhattan Project or with the challenge of beating the Soviets to the moon, but the race to find a substitute for the lacquer used to line food cans is heating up. The canning industry is frantically trying to find a replacement for the epoxy resin currently being used because of concerns that bisphenol A (BPA), the chemical we have already encountered as an "endocrine disruptor," may be leaching into the contents. BPA is combined with other components to form a polymer that keeps the metal from reacting with the food. Once the BPA has been incorporated into the polymer, it no longer has any hormonal effects, but there are always traces of unreacted BPA left over that can indeed leach out. Before exploring this issue, however, a bit of history is in order.

Napoleon, as many other generals before him, discovered that soldiers do not fight well on empty stomachs. And stomachs were often empty due to the difficulty of supplying food to massive traveling armies. So the emperor offered a prize of 12,000 francs, a healthy amount of money at the time, to anyone who could come up with a viable method of preserving food.

This challenge was taken up by Nicholas Appert, who, as the son of an innkeeper, had learned about brewing and pickling. He knew these "fermentation" methods could be halted by heat, and he began to wonder if food spoilage could also be stopped in this fashion. After all, it was clear that cooked food kept longer than fresh food, although eventually it too would spoil. Years of experimentation led Appert to make a critical discovery: if food was sealed in a glass jar and then heated, it would keep for a remarkably long time. Long enough to please Napoleon, at least, as he awarded the prize to Appert in 1809. The method clearly worked, although nobody at the time understood why. Bacteria were not identified as the cause of food spoilage until another famous Frenchman, Louis Pasteur, came along later in the century.

Appert's invention came to the attention of Peter Durand in England, who was troubled by the use of glass jars because they often broke. There had to be a better way! Why not a metal container? Iron was the first choice. But it would corrode, especially when exposed to acidic foods. A coating that would protect it from the air and contents had to be found. Tin, concluded Durand, would do the job! The metal had been known since antiquity and could easily be melted and applied to iron as a coating to make tin plate. And, most importantly, tin did not corrode. By 1818, the British Company Donkin and Hall was mass-producing food in tin cans. When Admiral Parry sailed to the Arctic Circle in 1824, he and his crew subsisted on canned food. One can of roast veal apparently was not consumed, because it turned up in a museum 114 years later. Inquisitive scientists opened it and decided to check the effectiveness of the canning process. They were not quite brave enough to try the veal themselves, but the rats and cats that had the pleasure of partaking of the 114-year-old feast not only survived, but thrived!

Although tin did not corrode, small amounts did dissolve, resulting in tainted food. This also meant the possibility of forming microscopic holes through which bacteria could enter and undermine the canning process. Aluminum eventually turned out to be more suitable for cans but still presented the problem of the metal interacting with the food. Chemists now stepped into the picture and found that an epoxy resin made by reacting bisphenol A with epichlorohydrin was excellent for providing a barrier that was stable under the high heat and pressure of sterilization, did not crack if the can was dented, and stood up well to the varying acidity of different foods.

Epoxy resins performed admirably, but cracks, figuratively speaking, began to appear in the early 1990s. By then, analytical techniques had been developed to detect extremely small amounts of BPA, and more importantly, the hormonal effects of this chemical were being demonstrated by its effects on the multiplication of cultured breast cancer cells. In 1995, researchers at the University of Granada in Spain investigated a number of canned foods and found estrogenic activity in peas, artichokes, green beans, corn, and mushrooms, but not in asparagus, palm hearts, peppers, or tomatoes. The authors pointed out that while an estrogenic effect was observed, it was far less than that observed for estradiol, the body's naturally occurring estrogen.

The significance of the estrogenic effect of canned foods is difficult to estimate given that, on top of the estrogen produced by the body, as we have previously seen, we are exposed to a wide variety of natural estrogenic compounds found in foods that include milk, chickpeas, soybeans, vegetable oils, cabbage, flaxseeds, and oats. It should also be noted that the concentration of pure bisphenol A required to produce maximum proliferation of breast cancer cells in the laboratory is 1,000-fold greater than for estradiol.

Even though no risk from traces of BPA in canned foods has been demonstrated, there is clamor for invoking the "precautionary principle," which aims to prevent harm even when the evidence is not fully in. For food companies, pleasing consumers is a high priority, whether consumers' demands are justified or not. So the race is on to find substitutes for epoxy resins. In some cases, for low-acid foods such as beans, plant extracts that harden into a resin have met with success. For other foods, companies are looking into various acrylics, polyesters, polyurethanes, and polyvinyl compounds. These do not match the performance of epoxy resin, nor is it clear that they have a better safety profile. Could we be trading in a perceived but unsubstantiated risk for a possible increased risk of food poisoning?

And one more thing: while you've been reading this little piece, hundreds of people have died from hunger, lack of clean water, poor sanitation, and a host of preventable diseases ranging from malaria to AIDS. By contrast, we have the luxury of worrying about traces of chemicals contaminating our ample food supply. A prescription for a dose of perspective is in order.

A Natural Conundrum

The Texas farmer was alarmed. Never before had he heard his cows bellow in this fashion. He rushed out to the pasture to see what was happening, only to be confronted by a horrific scene. The previously healthy animals were either staggering around or writhing on the ground. Eventually, fifteen of the eighteen cattle in the field perished. The veterinarian who conducted the necropsies concluded they had been poisoned by cyanide. The culprit, as it turned out, was the grass the cattle had been grazing on, a hybrid of two other grasses. It contained "cyanogens," compounds capable of releasing cyanide! So much for the facts. Now for some butchering of the same.

The world first heard about the cattle catastrophe from a CBS correspondent in Elgin, Texas, who, to the obvious delight of the anti-GMO crowd, filed a report under the headline "Genetically Modified Grass Linked to Cattle Deaths." Before long, a herd of bloggers and journalists piped in with alarmist stories about how "Genetically Modified Grass Kills Cattle by Producing Warfare Chemical Cyanide." But they were too quick to pull the trigger. They had not done their homework. The grass in question was not genetically modified; at least, not in the fashion that activists worry about. It was a hybrid grass, a product of traditional cross-breeding, and was in no way a novel product, having been around since 1983. It was, however, for some reason, in this particular pasture, producing an unusually large amount of cyanide.

Production of cyanide by plants is not a rare phenomenon. More than 2,600 cyanide-releasing species have been identified. Within the plant, the toxin is stored in an inactive form, bound to a sugar molecule, ready to be released as hydrogen cyanide upon reaction with an enzyme stored separately in the plant's tissues. The inactive compound and the enzyme are brought together when the plant is damaged, for example, when feasted upon by hungry insects. A whiff of cyanide and the insect is highly motivated to satisfy its hunger elsewhere. It seems these plants have evolved a mechanism to protect themselves from predators. And sometimes cattle, or even humans, can suffer the consequences as the plant unleashes its chemical defense system.

Perhaps the best example of the impact of cyanogens on humans is cassava, a plant we encountered earlier. It's a staple for millions of people in Africa, South America, and Asia. Like a potato, cassava's tuber-like roots can be boiled, fried, or processed into flour. The plant is easy to grow, is drought-resistant, and grows well without fertilizer. But it harbors a good dose of linamarin, a cyanogen. As discussed earlier, if not properly processed to rid it of cyanide, cassava can cripple or even kill. Thousands of children in Africa are victims of konzo, an irreversible paralysis of the legs caused by ingesting cyanide. Countless others suffer from headaches and dizziness due to low-grade cyanide poisoning. Drying, soaking in water, rinsing, and baking result in the cyanide being released into the air as hydrogen cyanide, but the process requires time. During periods of famine there is a tendency to shortcut procedures, and consumption of the improperly processed cassava can have tragic results.

If linamarin were eliminated from cassava, the time-consuming processing would not be needed. With the aid of genetic engineering, this is a distinct possibility. The gene that codes for the production of linamarin has been identified, and a method to silence it by interfering with the messenger RNA through which it sends out its information has been developed. Silencing cannot be total since some linamarin is needed by the plant to protect it from predators. But studies have shown that most of the linamarin is produced in the leaves from where it is ferried to the roots. Reducing leaf linamarin content by 40 percent still leaves plenty for protection and virtually eliminates the cyanide-producing compound from the roots. Further field trials are needed to ensure that inhibition of linimarin formation does not affect crop yields, since cyanide is a source of nitrogen and linamarin may be important in its transport from the leaves to the roots of the growing plant.

Yet another way of genetically modifying cassava may reduce its cyanide content. Cassava is quite low in protein but its content can be boosted by incorporating genes from sweet potatoes or corn that code for the production of a protein called zeolin. Enriching cassava with zeolin could save millions of children from potentially fatal protein-energy malnutrition. Furthermore, it turns out that cassava uses its natural supply of cyanide to produce the amino acids needed to build the new protein, thereby reducing the risk of cyanide toxicity. Again, further testing is required to ensure that the incorporation of the sweet potato or corn genes causes no untoward changes. But the possibility of saving human lives through genetic modification doesn't get as much play in the press as the demise of a few cows whose deaths were wrongly attributed to genetically modified grass by a bunch of bloggers and reporters when they came across a story that was just too juicy to check properly.

So, what did happen in that Texas field? The hybrid grass does contain dhurrin, a cyanogen. That is a fact. Why this grass that has long been used in cow pastures should all of a sudden produce lethal amounts of cyanide is not clear. Cyanide content is known to vary with growth, with the highest concentrations usually found in seedlings. Stress brought on by drought can lead to cyanide release, as can the use of nitrogen fertilizer at the wrong time, and the grass in the Texas pasture is known to have been heavily fertilized. Curiously, many other farms in the area grow the same kind of grass and have not experienced any problems. At this point, the only thing we can say for sure is that the cattle tragedy had nothing to do with genetically modified organisms. Obviously, nature can do plenty of damage without any help from humans.

Out of the Mouths of Babes

"Water Balz Toy Recalled." "35,000 Rubber Ducks in Santa, Reindeer Outfits Seized at Los Angeles Port." Not exactly the kind of headlines you like to see. What gives?

The ducks, it seems, were trying to duck regulations about the maximum amount of plasticizers called phthalates allowed in children's toys. But, contrary to the headline, they were not rubber ducks, they were polyvinylchloride (PVC) ducks. Had they really been rubber ducks, there would have been no issue, because rubber does not require plasticizers to make it pliable.

PVC is used in numerous items ranging from water pipes to shower curtains and, of course, toy duckies. It is a hard plastic but can be softened by blending in plasticizers. These do not react chemically with the polymer, but serve as sort of internal lubricants. Since they are not chemically bound, plasticizers, which can make up as much as 30 percent of the weight of the plastic, can leach out, albeit in small amounts. Nevertheless, this is an issue since some phthalates exhibit hormone-like properties. Since hormones can have biological effects in incredibly small amounts, there is an understandable concern about any chemical that may mimic the action of natural hormones in the body.

How do we know whether a chemical has hormone-like effects? Obviously it is not possible to purposely expose people to differing doses and watch for outcomes. Even if volunteers could be enlisted, and even if there were no ethical considerations, such a study would be practically impossible to carry out. "Endocrine disrupting effects" are subtle and may take decades to manifest. Evidence therefore comes not from randomized studies in people, but from the laboratory.

But what happens to cells in a petri dish can be very different from what happens in the body in the presence of thousands of other compounds that are either naturally produced or are introduced via eating, drinking, or breathing. Other evidence for the effect of chemicals can be obtained from observational studies that attempt to link exposure, often determined by blood or urine analysis, to measurable properties such as obesity, insulin resistance, or ano-genital distance. While associations can be found, they cannot prove a cause and effect relationship. Still, hormone-like behavior in the lab, or an indication of an effect in an observational study, does raise a red flag and does suggest adhering to reasonable precautions. Of course, opinions differ on what is meant by "reasonable precautions."

Some phthalates, but not all, have raised concern. That's why Health Canada has established specific limits. Diethylhexyl phthalate (DEHP), dibutyl phthalate (DBP), and benzylbutyl phthalate (BBP) cannot be present in toys or in child-care products to an extent of more than one gram per kilogram. Three others, diisononyl phthalate (DINP), diisodecyl phthalate (DIDP), and di-n-octyl phthalate (DNOP) are restricted to one gram per kilogram in any toy that infants may put in their mouth. California does not allow any phthalate in children's products to exceed one gram per kilogram. In Europe DEHP, DBP, and BBP are banned altogether in children's products and DINP, DIDP, and DNOP are not allowed in toys that children under age three might put in their mouth.

Regulations are regulations, and the ducks, imported from China, ran afoul of California law. Valued at over $18,000, the ducks were destined for destruction although their eventual fate wasn't clear. When it comes to PVC, incineration is not a good option because, at high temperatures, the plastic can yield the notoriously toxic dioxins. Whether a child chewing on a PVC duckie absorbs enough phthalate to represent a legitimate risk is a judgment call, but clearly avoiding exposure presents no risk.

The story of Water Balz is different because a risk has actually been documented. Until 2013, versions of polyacrylamide, a substance that has an amazing ability to absorb water, were available under names such as Water Balz, Growing Skulls, and Fabulous Flowers. In each case, the chemical came in the form of marble-sized colored balls, skulls, or flowers that expanded to roughly 400 times their original size when placed in water. They were advertised as appropriate for children over the age of three and included a warning about a "choking hazard." According to some of the ads, they were fun to grow, throw, or squish.

Why the recall took place in 2013 isn't clear given that the incident upon which it was based occurred in Texas in August 2011. That's when the parents of an eight-month-old infant girl took her to hospital because she was vomiting and had what seemed to be painful constipation. X-rays were unrevealing, but the baby's great-grandmother thought she had seen the child swallow a piece of candy. The candy turned out to be a super absorbent ball that was not actually sold as a toy but rather as a medium for holding flowers. "Crystal Soil" balls in a vase will slowly release water, allowing the flowers to flourish without regular watering. It really is neat to see flowers seemingly growing out of colored balls. Unfortunately, the brightly colored balls look like candy.

When the baby's condition worsened, doctors decided to operate. They found the water ball, which had expanded to a size big enough to block the lower portion of the small intestine. Clearly, polyacrylamide is not broken down by acid in the stomach. It was lucky that the surgery was performed because such intestinal blockages can be life-threatening. Since the balls were a popular item, the doctors in the Texas hospital decided to write up a case report for publication in the *Journal of Pediatrics* to alert other physicians and the public to the potential danger of swallowing superabsorbent balls.

The story was picked up by the press and received wide circulation. Reporters even managed to unearth three previous cases that required surgery because of blockages. Obviously, these balls may be safe to play with but not to eat. It is important to realize that children will put almost anything that they can get their hands on into their mouth. While the toys have been recalled, Crystal Soil is still widely available. And why shouldn't it be? It isn't marketed as a toy, and it can be used safely. As I've often said before, there are no safe or dangerous substances, only safe or dangerous ways to use them.

Cats, Calamities, and Static Cling

Have you ever wondered why, on some days, cats lick themselves more vigorously than on others? I suspect not. But their licking rate is indeed variable. And it just might have to do with the animal's fear of getting an electric shock. Unfortunately for felines, cat fur loses electrons very readily, and therein lies a problem. Anytime a cat rubs up against something — and they do a lot of rubbing up — electrons are transferred from the cat to the object, leaving the cat positively charged. When the animal now comes close to items that are good electrical conductors and therefore readily give up electrons, it is subjected to an electrifying experience. A spark, which is nothing more than a stream of electrons, can jump from the item to the cat. And then the cat jumps. Unless it has engaged in some prophylactic licking.

The buildup of static electricity is less likely when there is moisture in the air, due to a couple of factors. Water in the air makes the air more conductive, making for an easier dissipation of any charge that has built up. Furthermore, water molecules, being polar, also bind to the charged material. "Polar" means that within the molecule electrons are distributed in a fashion so as to make the oxygen atom slightly negative and the hydrogen atoms slightly positive. Cat fur being positively charged attracts the negative end of water molecules, which means the positive charge is partially neutralized, making the fur less attractive to any source of electrons. The risk of a spark is diminished. When humidity is low, the cat has to use saliva to moisten its fur to prevent being shocked. Since low humidity is usually associated with good weather, a cat licking itself with increased enthusiasm is a sign that rain is not likely. If you prefer not to use your cat as a barometer, a little spray with water will do the trick. But you may lose some affection.

Let's move on from licking cats to licking static cling. This too has to do with electron transfer. The tendency for such transfer is known as the triboelectric effect and the triboelectric series is a list of substances in order of their ability to lose or gain electrons. Substances at the top of the list tend to lose electrons readily, at the bottom, they are more likely to gain electrons.

The triboelectric effect was first described around 600 B.C., by the Greek mathematician Thales. Of course, there was no reference to electrons, which were not discovered until 2,500 years later by J.J. Thomson. Thales noted that light objects such as feathers were attracted to a chunk of amber that he had been polishing with a piece of fur. As we now understand, the rubbing transfers electrons from the fur to the amber, giving the latter a negative charge. When the negatively charged amber is brought close to a feather, it repels electrons from the feather's surface, making the surface positive. The attraction between the positive areas of the feather and the negative areas of the amber is an example of static cling.

A similar effect occurs when a plastic comb is run through hair. Since hair is above plastic in the triboelectric series, electrons are transferred from the hair to the comb, which can then pick up light objects just like the charged amber. Since the hair fibers have lost electrons, they become positively charged. Given that like charges repel each other, the result is the dreaded flyaway hair. The solution to this problem, as well as to that of static cling, is the neutralization of any charge that has built up by adding moisture to the surface. But spraying with water is usually not a practical solution.

In the case of hair, we turn to a conditioner composed of molecules that feature both a water-loving, or "hydrophilic," end, and a water-hating, or "hydrophobic," one. The hydrophobic end sticks to hair and the hydrophilic end attracts water, which then dissipates some of the charge on the hair. This same chemistry is used in commercial antistatic agents. A large variety of substances with both hydrophobic and hydrophilic properties are available, ranging from polyethyleneglycol esters to quaternary ammonium salts. The latter are also used in fabric softeners, thereby explaining why these also reduce static cling. Since fabric softeners are also lubricants, they further help to cut static buildup by reducing friction between surfaces.

Materials differ in their susceptibility to the buildup of an electric charge. The determining factor is the conductivity of the material, which to a large extent depends on its moisture content. Fibers such as silk, rayon, cotton, or wool have a relatively high moisture content and therefore charges are quickly dissipated. But synthetics such as polyester, polypropylene, and acrylics have a high surface resistance, meaning that electrons cannot readily move to neutralize a charge, particularly when humidity is low.

The latest technology to reduce the buildup of static involves the application of coatings that don't have much of a tendency to lose or gain electrons. An example is a special form of carbon known as a nanotube, in which carbon atoms are attached to each other to form a cylindrical molecule. These molecules aggregate together to form nanoparticles less than 100 nanometers in diameter. These nanoparticles form a strong bond to fibers, don't lose or gain electrons, and are also excellent lubricants. Fabrics coated with nanoparticles also feel soft, resist stains, and dry readily.

Had Frank Clewer in Australia been wearing antistatic garments, he would not have caused the stir that he did back in 2005. But he was wearing wool and nylon, both of which are high on the triboelectric series, meaning they readily assume a positive charge. When Mr. Clewer walked into a building for a job interview, he set the carpet on fire by causing sparks as electrons jumped from the synthetic material toward his positively charged clothing. The heat generated was enough to ignite the carpet, necessitating the evacuation of the building. There have been no reports of cats sparking such calamities. So when you see your cat licking himself with great gusto, he's not only protecting himself from electric shocks, he's protecting your home from a fire. By the way, Mr. Clewer didn't get the job.

Now perhaps you understand why taking off polyester pants or acrylic sweaters in the dark can cause sparks. You're just watching electrons jumping from negative to positive surfaces. If you don't want this electrifying experience, apply some antistatic spray, or just spray with a little water. Or, just take off the clothes more slowly.

Chemistry in the Spotlight — for a Tragic Reason

The eyes of the academic chemistry community have been riveted on a courtroom in Los Angeles, where UCLA chemistry professor Patrick Harran stands accused of "willfully violating occupational safety and health standards and causing the death of a young technician in his laboratory." Many professors are following the trial with trepidation, mindful of the possibility that they could be the ones facing the music in that courtroom. At the time of writing, there is only one certainty about this evolving drama: it is tragic for everyone involved. A young woman with great promise for the future is gone, her parents' lives now dominated by weekly visits to the cemetery. A distinguished professor's life is shredded as he faces a possible prison sentence.

Dr. Patrick Harran is a researcher and teacher with a stellar record of awards and publications. One of his interests is appetite-suppressant drugs and it was in this connection that twenty-three-year-old Sheri Sangji was performing an experiment in his laboratory on December 29, 2008. Sangji was equipped with an undergraduate degree in chemistry and had been working for a pharmaceutical company when she was hired by Harran as a research assistant. The particular reaction she was working on required the use of tert-butyllithium, a notoriously pyrophoric compound, meaning that it bursts into flames on contact with air. Obviously, it requires special handling.

As I mentioned earlier, in the scientific community, we are fond of saying that there are no safe or dangerous substances, only safe or dangerous ways to use them. And so it is with tert-butyllithium. If the detailed instructions provided by the manufacturer are properly followed, there should be no problem. One method of transfer uses a syringe, and it was the one followed by Sangji, but unfortunately not according to the instructions. The bottle wasn't clamped, the syringe was too small for the amount being dispensed, and the needle used was too short, requiring the bottle to be tilted. Although the exact details are murky, Sangji accidentally pulled the plunger out of the syringe, allowing the liquid to escape. It instantly burst into flames, igniting her nitrile gloves and synthetic sweater. She was not wearing a lab coat. In her panic, Sheri did not run toward the safety shower in the lab, and by the time a lab mate managed to extinguish the flames, she had suffered extensive burns. In spite of care at one of the best burn centers in the U.S., Sheri Sangji passed away three weeks later.

An investigation was immediately launched not only by the university, but also by California's Occupational Safety and Health Administration (OSHA). Investigator Brian Baudendistel carried out extensive interviews with everyone connected with the case and put together a report accusing both the university and Harran with laxity in implementing proper safety procedures. He urged that both be charged with involuntary manslaughter. Baudendistel concluded that the professor had not discussed the specific risks of working with tert-butyllithium with Sangji and had not enforced the wearing of lab coats with enough vigor. He had, according to the report, "permitted Victim Sangji to work in a manner that knowingly caused her to be exposed to a serious and foreseeable risk of serious injury or death." Furthermore, Baudendistel discovered that UCLA had received previous warnings about its safety standards and that Dr. Harran's lab had been cited by UCLA safety inspectors for violations, including failure to enforce the wearing of protective gear.

The district attorney's office took two years to scrutinize the OSHA report. There would be no manslaughter charge, as the OSHA investigator had requested. But after considering that there had been other recent accidents in UCLA labs that had not been properly dealt with, a decision was made to prosecute the university and Professor Harran for violating occupational safety and health standards. A warrant was issued for Harran, who was on vacation at the time. As soon as he returned, he was arrested and a trial date was set.

In July 2012, the prosecutor dropped felony charges against

the university in return for a guarantee that a number of safety measures would be instituted. All professors and laboratory personnel would henceforth be required to complete a lab-safety training program. Standard operating procedures must be written and reviewed by experienced, qualified personnel, and these must be followed rigorously. Anyone not wearing proper personal protective equipment must be removed from the lab and the incident documented. Regular chemical safety inspections have to be conducted and accepted procedures for the safe use of pyrophoric liquids must be followed. All occupational injuries and illnesses must be reported to California's Office for Occupational Safety and Health Administration.

Charges against Professor Harran were left to stand with a possible sentence of four and a half years in prison hanging over his head. Surprisingly, as we were all gearing up to follow reports of the trial, the case suddenly took a new twist. Harran's defense attorney introduced a motion to dismiss the charges because, he claims, evidence gathered by OSHA investigator Baudendistel should be inadmissible. Why? Because Baudendistel was convicted of having helped set up the murder of a drug dealer back in 1985. Since he was a juvenile at the time, records have been sealed. The defense attorney claims that a man with such a history is not credible, but Baudendistel insists that they have the wrong man, apparently in spite of some fingerprint evidence. It seems to me, though, that this case should be decided based on the evidence and not dismissed on account of some irrelevant technicality.

As one might expect, the internet is abuzz with thoughtful as well as inane commentaries on this extraordinary legal case. Some lay the blame on Sangji, claiming that she should have followed proper procedures and should have been wearing a lab coat, which in this case could have been life saving. Others accuse Dr. Harran of murder for not properly supervising a dangerous reaction. But many chemists realize that neither UCLA nor Dr. Harran are unique examples of negligence in terms of safety, and recognize that their own closets may harbor skeletons. We can all hope that this sad case will cause institutions and individuals to reflect on their safety procedures and make improvements where needed. Indeed, according to a U.S. federal investigation, there have been about 120 serious lab accidents in universities between 2001 and 2011. Perhaps Sheri Sangji's tragic death will help reduce this toll.

Just Ironing Things Out

You may find this surprising, but I don't mind ironing. Unlike giving a lecture, writing a column, or appearing on TV or radio, you get immediate gratification. You see the results of your efforts. Wrinkles disappear. I suspect, however, that not everyone shares my enthusiasm for this task. The textile industry realizes this as well and has responded by producing a variety of "durable press" fabrics that can withstand wrinkles. However, withstanding allegations of toxicity is more of a challenge. In this case, the hullabaloo is about formaldehyde, the chemical used to fashion garments that can come straight out of the washing machine and sidestep the ironing board.

Wrinkling is a direct consequence of the molecular structure of cellulose, the main component of cotton. This polymer is made up of repeating units of glucose, but the important feature of cellulose, as far as wrinkling is concerned, is that adjacent molecules can form weak associations with each other. These hydrogen bonds are responsible for maintaining the shape of the fabric. But when cotton is moistened, water molecules insert themselves between the long chains of cellulose, cleaving the hydrogen bonds. The cellulose molecules can now move relative to one another, and as the fabric dries and the water molecules evaporate, the hydrogen bonds reform to hold the fabric in its new shape, which is usually wrinkled. Another factor in wrinkling is the thickness of the cotton fibers. Fabric woven with very fine cotton thread will crease less than fabric made with a coarser thread.

Heat can also disrupt hydrogen bonds, which explains how ironing works. The weight of the iron flattens the fabric and the novel shape is then retained as the material cools. Ironing with steam is especially effective because the added water molecules serve as an internal lubricant, breaking hydrogen bonds and allowing cellulose molecules to slide past each other. As heat is applied, the water evaporates, hydrogen bonds reform, and we have a smoothened fabric. At least, smooth until it gets moist from perspiration.

The earliest attempts to reduce wrinkling made use of starch, possibly as early as 800 B.C. Like cellulose, starch is made of repeating units of glucose, and the two substances have an affinity for each other. A starch solution readily penetrates into cotton fibers and when the fabric dries, the molecules of starch bind to each other, forming a hard network and stiffening the fiber. An analogy would be a canvas water hose that can easily be folded when empty, but becomes stiff when filled with water. The problem with starch, though, is that it comes out in the wash and has to be applied each time.

The early part of the twentieth century saw the introduction of a number of synthetic resins such as urea-formaldehyde, which presented textile manufacturers with an opportunity to "stuff" fibers with a substance that would not wash out as readily as starch. It worked reasonably well, but when the chemical structure of cellulose became clarified in the late 1920s, chemists came up with another idea. Given that wrinkling is caused by the movement of the cellulose molecules relative to each other, why not find a way to form bonds between cellulose molecules that are not as easily disrupted by moisture as hydrogen bonds?

Formaldehyde was just the molecule for this task. It readily reacts with the hydroxyl groups on cellulose to form cross-links, much like the rungs of a ladder. The idea is to form the fabric into the desired shape and then treat it with formaldehyde to retain that shape. Manchester textile manufacturer Tootal Broadhurst Lee was the first company to use this process commercially, back in the 1930s, producing the world's first "wrinkle-free" garment, the Tootal tie. But it wasn't long before a wrinkle appeared in the novel technology. The garments released the irritating odor of formaldehyde. Not only was this smell unpleasant, but formaldehyde was also responsible for the allergic contact dermatitis reactions that emerged with the use of the durable press fabrics. Later, there would be increased concern about formaldehyde treatments as studies began to show that, at least in animals, the chemical was a carcinogen.

By the 1990s, significant improvement had been made in technology to reduce formaldehyde release from treated fabrics. During manufacture, the fabric is subjected to huge rollers that squeeze out excess formaldehyde followed by heat treatment in an oven to cure the formaldehyde and prevent it from being released. Another method that requires less formaldehyde, known as vapor phase technology, involves hanging moistened clothing in an airtight chamber and treating it with a gaseous mixture of formaldehyde and sulphur dioxide to form the required cross-links.

A number of other cross-linking agents have also been introduced, with the most widely used ones being dimethylol dihydroxyethyleneurea (DMDHEU) and ethylene urea-melamine formaldehyde (EUMF). While effective at producing wrinkle-free fabrics, they, like formaldehyde, can still cause allergic reactions. Exposure to permanent-press fabrics should always be considered as a possible cause of dermatitis that has no obvious trigger.

Although the amount of formaldehyde released from permanent press fabrics is unlikely to affect health except for rare cases of allergic dermatitis, manufacturers have made determined efforts to reduce formaldehyde exposure. The textile industry uses a Sealed Jar Test to measure the amount of formaldehyde released from one gram of fabric under controlled conditions. Since the early days of durable press, the amount released has been decreased by a factor of ten. Cross-linking agents that do not release formaldehyde at all, such as dimethylurea glyoxal (DMUG), have also been developed but they cost more and do not perform quite as well.

The advent, in the 1960s, of polyester fabrics with a greatly reduced tendency to wrinkle took some of the pressure off producing wrinkle-resistant cotton fabrics. But since not everyone likes the feel of polyester, even when blended with cotton, the prospect of producing a truly permanent-press cotton fabric still looms in front of manufacturers' eyes. They're getting there, though. I recently bought some shirts that barely need the touch of an iron. And as far as formaldehyde exposure from such fabrics goes, well, I think there are far bigger wrinkles in the fabric of life to worry about.

Meat Production: A Smelly Business

Meat production stinks. And I'm not referring to worries about bacterial contamination. I mean it literally stinks. Here's the story. Hold your nose.

We crave meat. To satisfy our hunger, American feedlots ready some 150 million cattle for slaughter every year, while in Canada, the number is around 3.5 million. And those cattle produce more than meat. A single steer can crank out up to 30 kilograms, or 66 pounds, of manure and urine every day, and some feedlots house over 40,000 animals! That means more than a thousand tons of pee and poo have to be dealt with in some fashion every day! Obviously, all that waste creates a massive odor problem caused by a huge array of smelly compounds that are released as a result of bacterial action on manure. Most of these compounds have been identified, with indole, skatole, and dimethyl sulphide, which are all also present in human excreta, having been found to be particularly noxious.

Any effective odor control process has to destroy these compounds one way or another. Oxidation, as the term implies, requires reaction with oxygen, the simplest example of which is the process we know as combustion. Just think of how natural gas, basically methane, combines with oxygen as it burns to yield carbon dioxide and water. Other organic compounds, such as the variety of aldehydes, ketones, acids, esters, and amines released from manure can also "burn," but of course, setting fire to feedlots is not a solution. There are, however, oxidizing agents that have the ability to transfer oxygen to a variety of other compounds in what can be called a "cold" combustion process. Potassium permanganate is one of these, and a solution can be easily sprayed on the ground to control the odor of manure. Spraying with a 1 percent solution at the rate of about 8 kilograms of permanganate per acre three times a year is very effective at reducing the odors from cattle feedlot operations. If animals are raised in an enclosed space, ozone, a very strong oxidizing agent, can be used to reduce odors. In such feedlots, air can also be collected and passed through beds of activated carbon that adsorb the smelly components, but this requires periodic replacement of the carbon, which has a relatively short adsorbent life, making the process expensive.

Methods of altering feed in order to reduce odors have also been investigated. Humic acid is the black spongy matter present in soils that is a mix of compounds formed when dead plant matter undergoes microbiological degradation. When incorporated into animal feed, it can lead to substantial reduction in the odor of manure.

Of course, cattle feedlots are not the only place where odor problems are encountered. Dairy, poultry, hog, and sheep farms also battle the problem. And it's a big problem. The total number of animals raised to meet human demands is stunning. At any given time, North American farms are estimated to house some 102 million cattle, 82 million hogs, 7 million sheep, and close to 600 million chickens. Just imagine the total amount of smelly compounds produced by all that manure. But smell is not the only issue. High odorant concentrations can kill animals, and the effects of long-term exposure of sub-lethal amounts are unknown.

Then there is the problem of workers being affected by gases and particulate matter released from manure. Hydrogen sulphide, responsible for the classic odor of rotten eggs, is one of the components of manure smell and is highly toxic. A number of farm workers have died while attempting to clean manure pits after being overcome by hydrogen sulphide. There is also the problem of developing a lung disease known as allergic alveolitis after long-term exposure to particulate matter containing a variety of antigens (compounds capable of causing allergies), particularly in chicken droppings. Even aside from health issues, the revolting smells released by animal-raising operations can be a nuisance to anyone living downwind.

Smells are also a major problem in animal processing and rendering operations. Fish meal processing produces trimethylamine and putrescine, both with terrifying smells. Rendering of beef offal yields many odorous compounds, and processing of feathers generates the likes of acrolein, acetaldehyde, methyl mercaptan, diethylamine, n-propylamine, ammonia, and hydrogen sulphide. Many of these can be eliminated by circulating the air inside the facility through "chemical scrubber" solutions: a solution of limewater can be used to remove ammonia; hydrogen sulphide can be removed with potassium permanganate; aldehyde smells can be removed using a sodium bisulphite solution; sodium carbonate can neutralize acids; and calcium hypochlorite is a powerful oxidizing agent.

Environmental issues also crop up. Ammonia released from animal excreta can be absorbed by nearby bodies of water, where it can stimulate the growth of algae, which in turn uses up the dissolved oxygen content of the water, depriving fish of oxygen. Basically, animal production facilities are a major source of air pollution as well as of water pollution from feedlot runoff.

And then there is the problem of methane production. Ruminant animals, such as cattle, sheep, buffalo, and goats have a digestive system that can convert otherwise unusable plant materials into nutritious food and fiber. But this same helpful digestive system also produces methane, a gas that has no smell but is a potent greenhouse gas that plays a role in global warming. Livestock production systems can also emit other greenhouse gases such as nitrous oxide and carbon dioxide. Globally, ruminant livestock produce about 80 million metric tons of methane annually, accounting for about 28 percent of global methane emissions from human-related activities. An adult cow may be a very small source by itself, emitting only 80 to 110 kilograms of methane a year, but with about 1.2 billion large ruminants in the world, they constitute one of the largest methane sources.

There is yet another issue: the raising of animals requires huge amounts of water. It's not only the water they drink, it's also all the water used to grow the crops they eat. So, there's no question about it, raising animals is not an environmentally friendly process, and is not an efficient use of crops or water. So why do we do it? I think the simple answer is that most of us like the way they taste. And when environmental issues raise a stink, we just hold our noses.

Reflecting on the History of Mirrors

Mirror, mirror on the wall, what neat chemistry is behind it all? Take a moment to reflect on what life would be like without mirrors. The cosmetic industry might not exist, dentistry would be a challenge, and driving would be a nightmare.

Our ancestors probably first took note of a reflected image as they gazed into the still waters of a pond. But that sort of mirror wasn't very portable. Polished reflective stones dating back to 6000 B.C. have been found in Turkey, and by 3500 B.C., the Sumerians in Mesopotamia had developed methods of polishing brass with sand until it developed a reflective surface. The ancient Egyptians, Israelites, Greeks, and Romans all gazed at their images in mirrors made from polished copper, bronze, tin, silver, or gold. Fragments of glass coated with tin, found by archeologists in Roman ruins, may represent the first glass mirrors. By 500 A.D., Chinese craftsmen had developed methods to coat glass with an amalgam of silver and mercury, but it took another thousand years until high-quality mirrors were produced in Venice by using a tin-mercury amalgam. The technique involved lining a marble table with tin foil and then covering it with mercury. Slowly, either by hand or with a special brush, the mercury was worked into the tin to form a malleable amalgam that was then smoothed onto a glass plate. Little surprise that mirror makers often died early from mercury poisoning!

"Poison the treacherous mirror makers!" That, at least according to some accounts, was the edict delivered in 1665 by the infamous Venetian Council of Ten. The victims were to be the skilled glassblowers who had been enticed to France by Jean-Baptiste Colbert, Louis XIV's finance minister, to head up the newly established Manufacture Royale de Glaces de Miroirs. At the time, Venice was the center of the glass and mirror trade, with artisans jealously guarding their secrets. Colbert, however, was determined to achieve French self-sufficiency in manufacturing, especially when it came to the furnishings of the Hall of Mirrors, destined to become the crowning glory of the Palace of Versailles.

Income from the export of glass and mirrors was a huge boost to the Venetian economy. Understandably, the Council of Ten, a tribunal charged with looking after the political, moral, and financial welfare of the state, was disturbed by the prospect of the French acquiring the Venetian technology. Poisoning, mostly with arsenic, belladonna, aconite, hellebore, or strychnine, had evolved into an effective way of dispatching undesirables in Renaissance Italy, and it became the council's preferred method of dealing with enemies.

Great secrecy was maintained, but the council did keep private records in the *Secreto Secretissima*, a book now on display in a Venetian museum. One entry describes how John of Ragusa, a Franciscan brother, offered his poisoning talents to the council, with a curious price list. The Great Sultan would be disposed of for 500 ducats and the King of Spain for 150. The Pope was a bargain at 100 ducats. It isn't clear what the fee for treacherous mirror craftsmen would have been, but perhaps it was too high because history does not record any unusual deaths among the imported Venetians at the Manufacture Royale. And it seems the French did manage to learn the secrets of mirror production from the Venetian expatriates because by 1672, the importing of glass or mirrors into France was forbidden. To this day, the French are proud of the fact that Louis and all of his mistresses were able to admire themselves in Versailles's fabulous mirrors, all of which were made in France.

What, then, was the secret of Venetian mirror-making? To be sure, the Venetians did not invent mirrors. Glassblowers discovered that introducing molten antimony, tin, or lead into a freshly blown globe, and swirling the metal as it cooled, led to a thin deposit of the metal on the glass. Cutting the globe into pieces then produced mirrors of good quality, but the image of course was somewhat distorted due to the convex shape.

This is where the Venetian glassblowers come into the picture. They revolutionized mirror-making by first finding a way to cut open a globe of blown glass and flatten it, and more importantly, by managing to coat it with an alloy of tin and mercury. The process was called silvering because the coated glass had a silver color, but of course this was a misnomer as no silver was involved. A piece of tin is pounded until it is paper-thin, and then covered with mercury. The mercury seeps into the tin to form a homogeneous amalgam. At this point, a sheet of glass is placed on top and weighted down to squeeze out excess mercury. Heat from a fire or the sun then bakes the metallic coating onto the glass, and presto, you have a mirror! This is the technology that was passed on to the French, the results of which we see today in the grand Hall of Mirrors. While there's no record of the Council of Ten making good on its poisoning threat, there is a good chance that many of the mirror-makers were indeed poisoned. Not by arsenic or belladonna, but by the ever-present mercury fumes!

It would have been such a mercury-coated mirror that set the evil queen in *Snow White* on her wicked way by declaring that she was not the fairest in the land. The Grimm Brothers' classic story was published in 1812, well before noted German chemist Justus von Liebig's discovery that made high-quality, safe mirrors available. In Liebig's process, instead of a mercury amalgam, a layer of metallic silver was deposited on the glass directly by means of a chemical reaction between glucose and a solution of silver nitrate in ammonia. The silver nitrate was reduced to metallic silver as glucose was oxidized to gluconic acid. Liebig carried out his initial experiments in 1835, but a commercial process for making mirrors by his method was not developed until the late 1850s. By the time that Alice went through the looking glass in 1871, mirrors had become commonplace, thanks to Liebig's procedure. Other scientists, such as the English chemist Thomas Drayton, had also experimented with depositing silver from a silver nitrate solution, but were unable to make the process commercially viable.

Liebig was originally driven by a desire to protect mirror-makers from the dangers of exposure to mercury. But he also became interested in producing higher-quality mirrors when his friend, Munich physicist and astronomer Carl von Steinheil, expressed a need to improve the metallic mirrors being used in telescopes. It was at this point that Liebig discovered that the addition of a little copper to the silver nitrate solution resulted in the deposition of a blemish-free layer of silver. No doubt he was happy to help out his friend, but truth be told, Liebig was not averse to capitalizing financially on his discovery. After all, at the time, professors of chemistry, no matter how accomplished, were not well paid. Liebig eventually took out patents and made some money, but his process could not compete economically with the mercury amalgam mirror industry. After Liebig's death, when he unfortunately could no longer benefit, the German government passed legislation that prohibited the use of mercury in mirrors.

Although Liebig's process produced high-quality reflective surfaces, it also produced problems. One of the byproducts of the reaction was ammonium nitrate, an explosive! In fact, if a residue of this substance was left on the mirror, the mirror could crack at the slightest disturbance. This possibly is the origin of the notion that an ugly face can crack a mirror.

Like silver anywhere else, the backing on a Liebig-type mirror can tarnish. Reaction of the silver with sulfur compounds in the air can result in the formation of dark, non-reflective silver sulfide. This is usually not a problem as long as the silver is deposited onto the glass in an airtight fashion. But if there are air pockets between the glass and the silver, water can seep in, and since this speeds up the tarnishing reaction, an unsightly black edge can develop. The preventive technique is to wipe off excess water from the edge immediately after cleaning a silvered mirror.

But modern mirrors are rarely silvered. They're generally made by applying a coating of aluminum instead of silver to the glass. It is cheaper and reflects very well. The glass to be coated is masked on one side and is then suspended in a vacuum chamber in which powdered aluminum is heated until it vaporizes. Since the glass is cooler than the vaporized aluminum, the metal condenses on the surface to form a smooth reflective coating.

Normal mirrors have the metallic coating on the back of the glass, but a coating on the front can produce a "two-way mirror." A one-way mirror reflects all light, but when the coating is on the front, some light passes through. Such two-way mirrors are the kind used for spying. If installed in a brightly lit room with a darker room behind the mirror, an observer in the darkened room can see through the mirror but cannot be seen from the other side. That's exactly what you want in a police lineup. And how can you tell if a mirror is one-way or two-way? Easy. Just touch your finger to the mirror. If the coating is on the front — that is, if it is a two-way mirror — the finger and reflection will touch. In a one-way mirror, where the coating is on the back of the glass, the finger and the reflection will not touch.

There's one last point to remember: vampires do not have mirror images. That makes it difficult to tell if one is sneaking up on you while you are admiring yourself in a mirror. If you want protection, invest in a high-quality silvered, not aluminized, mirror. Vampires are said to be allergic to silver.

Dry Ice — It's Sublime!

"Double, double, toil, and trouble; / Fire burn, and cauldron bubble!" But how do you make that cauldron bubble on stage? That's what the stage manager of a local production of *Macbeth* wanted to know. It wasn't too hard to answer that one. It just takes a little chemical witchcraft in the form of dry ice!

When ordinary ice melts, it transforms into water. But dry ice isn't ice at all. It is solid carbon dioxide. When it warms up, no liquid forms. This "ice" remains totally dry as it slowly appears to vanish. "Appears" is an important term because dry ice does not break nature's fundamental law that matter cannot be created or destroyed, it can only change from one form to another. In this case, solid carbon dioxide changes directly from a solid to a gas without going through the liquid state. This process is known as sublimation, and for dry ice occurs at the very low temperature of −78.5°C, or −109.3°F. Since at normal atmospheric pressure, dry ice cannot exist as a solid above this temperature, it can serve as an excellent cooling agent when refrigeration is not available. That's why food and biological samples that need to be kept cold can be packaged with dry ice for shipping.

Dry ice is manufactured from carbon dioxide gas. Gases can be turned into liquids by applying pressure, and this is the first step in the production of dry ice. When the pressure is decreased, the liquid immediately begins to revert to a gas. The heat needed to convert the liquid to a gas is drawn from the liquid carbon dioxide, which then experiences a quick drop in temperature and freezes into a solid. This change can be readily demonstrated in the laboratory by placing a few pieces of dry ice into a specially made strong test tube equipped with a screw cap. When the tube is sealed and the temperature begins to rise, the dry ice sublimes and the gas that forms increases the pressure inside the tube. With the increased pressure, the solid carbon dioxide is seen to change into a liquid. Unscrewing the top immediately decreases the pressure; the liquid begins to evaporate, the temperature drops, and in the wink of an eye, the liquid carbon dioxide converts into dry ice.

The pressure produced in the sealed tube by the subliming carbon dioxide can quickly reach several atmospheres. If the tube is not strong enough to withstand the pressure, then instead of a neat science experiment, we have a bomb! This possibility unfortunately has an attraction both for adventurous students and for aspiring terrorists. Sealing dry ice into plastic bottles can result in quite an explosion, sometimes even before the bottle leaves the hand, causing serious injuries. In many jurisdictions, any attempt to seal dry ice in a bottle is a criminal offence and can lead to imprisonment.

Next question is where the original carbon dioxide gas comes from. Carbon dioxide is present in the air, where it serves as the key to all life on Earth, as it is the raw material needed for photosynthesis, the process by which plants make glucose. But isolation of carbon dioxide from air is not a practical process. However, huge amounts of carbon dioxide are produced as a byproduct when natural gas (mostly methane) reacts with water to form hydrogen. This is one of the world's most important industrial processes since hydrogen is needed for the production of ammonia, which in turn is converted into fertilizer. Without synthetic fertilizer, the world's demands for food could not be met.

Both dry ice and liquid carbon dioxide have a number of uses besides providing non-mechanical cooling. Liquid CO_2 is an excellent environmentally friendly solvent and can be used, for example, to remove caffeine from coffee beans. It is also an alternative to potentially toxic organic solvents in the dry cleaning industry. Solid carbon dioxide can be used instead of sand for blast cleaning, the advantage being that it leaves no residue as it sublimes. But perhaps the most familiar application is in the production of theatrical special effects. When dry ice is dropped into water, it quickly warms up and changes into a gas, forming impressive bubbles. That is how you make the witches' cauldron bubble! As a bonus, a mysterious mist cascades to the floor.

The carbon dioxide gas that escapes from the container is invisible, but still very cold. So cold, in fact, that it causes moisture in the air to condense into a liquid. The "fog" that forms is actually composed of tiny droplets of water, just as in the case of real fog. Carbon dioxide is also heavier than air, which is why theatrical fog spreads over the floor instead of rising upwards.

While carbon dioxide is not toxic, it can kill. How? By suffocation through the displacement of oxygen. In a classic case in 1940, five longshoremen who had bedded down in the cargo bay of a ship transporting cherries to New York from Michigan died during their sleep when the dry ice used to keep the cherries cool sublimed and displaced the air, settling over the men like an invisible asphyxiating blanket. Such stories have given rise to speculations about committing the perfect murder by sneaking dry ice into a sleeping victim's room. After the dry ice sublimed and did its dirty work, there would be no trace of a murder weapon left. Not totally impossible, but unlikely given the requirement for a virtually sealed room.

The witches onstage have nothing to worry about. At least, not from the carbon dioxide. But then there is the curse of *Macbeth*. According to one superstition, the original propmaster could not find a cauldron and stole one from a real witches' coven. The witches are said to have cursed the play in retaliation, which is why productions of *Macbeth* are supposedly plagued by all sorts of calamities. The production I consulted on was performed outdoors. As soon as the dry ice was dropped into the water, it started to rain cats and dogs. Hmmm.

The Cuddle Chemical Versus Personality

Why do people cuddle? According to a slew of media reports, the answer lies in oxytocin, dubbed either the "cuddle chemical" or the "love hormone." English pharmacologist and neurophysiologist Sir Henry Hallett Dale was the first to isolate oxytocin from pituitary extract in 1921, eventually receiving the 1936 Nobel Prize in medicine and physiology for the discovery. Dale's work on oxytocin was prompted by his interest in the ergot fungus, extracts of which had a long history of use for stimulating the contractions of the pregnant uterus. The question was why.

In 1909, Dale had prepared an extract from the posterior lobe of the pituitary and showed that it caused contractions of the uterus when injected into a pregnant cat. As a result, pituitary extract replaced ergot preparations as the prime method to induce labor. Dale hypothesized that the reason ergot stimulated contractions was that it contained some component that resembled a substance that women in labor produced naturally. But what was that component? Dale wasn't successful in isolating the active ingredient from ergot, but eventually did manage to isolate oxytocin from pituitary glands, deriving the name of the newly discovered substance from the Greek for "quick birth." Ergometrine, the active ingredient in ergot, was finally isolated and identified in 1935 by one of Dale's former colleagues, Harold Ward Dudley.

The concept of hormones as chemical messengers had been introduced in the first decade of the twentieth century by British researchers Bayliss and Starling, who had found that entry of food into the small intestine caused the mucosa to release a chemical that then traveled through the bloodstream to the pancreas, where it stimulated the release of pancreatic juices and bile needed for digestion. Secretin became the first hormone to be isolated, the term "hormone" deriving from the Greek "to excite." This was the first demonstration that chemical action could be transmitted to remote parts of an organism without involvement of the nervous system. The search was now on for other such messengers.

The brain's pea-sized pituitary gland was a candidate for the production of chemical messengers because autopsies had revealed that people who grew to be "giants" had enlarged pituitaries. It seemed possible that the enlarged gland was sending some sort of message to the rest of the body, signaling it to grow. In 1912, Henry Cushing would postulate the existence of a "hormone of growth" produced by the pituitary. He turned out to be correct. But even before this, as we saw, Henry Dale had been working on pituitary extracts, finally isolating oxytocin in 1921. The compound's exact chemical structure, a string of nine amino acids, commonly referred to as a polypeptide, was not determined until 1953. That same year, oxytocin became the first polypeptide hormone to be synthesized in the laboratory and the synthetic version quickly replaced pituitary extract as the drug of choice in the delivery room.

Today, synthetic oxytocin is widely used to induce labor — too widely, according to some. The concern has nothing to do with the fact that a synthetic version is used; oxytocin is oxytocin, whether made in the lab or in the body. Rather, the issue is over the amount that is administered, which does not exactly mimic natural production, and that in some cases can cause contractions that are too severe. Perhaps more importantly, it is becoming clear that oxytocin affects behavior in various ways and can conceivably produce effects in exposed babies down the road.

The behavioral connection first emerged from an investigation of the love life of the prairie vole. These mammals are unusual in that they are monogamous. What makes them so? It seems their love is rooted in the release of oxytocin. When a couple engages in sex for the first time, oxytocin is released, somehow formalizing the union. From then on, the voles only have eyes for each other. Blocking the release of oxytocin with an oxytocin antagonist, which is basically a modified form of oxytocin, results in one-night stands. Should prairie voles be injected with oxytocin, they will search for a partner, and even if they are prevented from having sex, they will continue to stay with the chosen partner.

On the other hand, a close relative, the montane vole, is immune to the effects of oxytocin, apparently having no receptors for the chemical. Other mammals do have receptors, even though they may not be monogamous. Sheep, for example, reject their young if treated with oxytocin antagonists, and female rats injected with oxytocin will nurture another's pups as if they were their own. In rats, injection of oxytocin into the cerebrospinal fluid causes spontaneous erections, and there is some evidence that the hormone also plays a role in the female's willingness to bend to the male's desires. Some marketers of oxytocin highlight this effect and insinuate, without any evidence, that the chemical might have such uplifting effects in humans as well.

Humans obviously do have receptors for oxytocin, otherwise the chemical would have no biological activity. Aside from uterine contractions, the chemical appears to play a role in stimulating bonding between mother and child. Studies have also shown that humans who sniff oxytocin via a nasal spray become more trusting. In one interesting study, volunteers treated with oxytocin invested more money in a questionable business venture than those treated with a placebo, even when they were told that there was no guarantee that the trustee was trustworthy.

Other studies have linked exposure to oxytocin with reduced social anxiety, but there have also been some disturbing observations with existing biases being strengthened when oxytocin was inhaled. Because of publicity given to preliminary data about improved social connections, it comes as no surprise that oxytocin nasal spray is being promoted on the web as a treatment for autism. There is some evidence of minor benefits, but long-term risks are unknown. As far as cuddling or falling in love, or some sort of aphrodisiac effect, unfortunately there's no evidence. Experiments with human couples sniffing oxytocin have not shown any increased tendency to fall in love or even to cuddle. Better to rely on personality than a nasal spray.

Lighter, Brighter, Safer!

"Faster, higher, stronger!" That's what Olympic athletes strive for. But for designers of the Olympic torch, the motto is "lighter, brighter, safer." The torch is perhaps the most significant symbol of the Olympics, but few realize the effort that goes into its design both in terms of aesthetics and science. The torch has to be light, has to burn with a bright flame, and cannot present a health risk to those carrying it.

The symbol of the flame goes all the way back to the first Olympic Games in 776 B.C. The ancient Greeks revered fire, appreciating its role in the progress of civilization. According to mythology, fire was of divine origin, stolen from Zeus, the "Father of Gods and men," by the Titan Prometheus, who then introduced it to humans. Since the original Games honored Zeus and the other gods who lived on Mount Olympus, a flame became a natural symbol of the Olympics. Hera, who was both wife and sister to Zeus (let's not even go there), was one of the gods honored, and it was on her altar that the Olympic flame was first kindled by using a mirror to focus sunlight. Runners then relayed the flame to the site of the games.

The ancient Olympic Games were celebrated for about a thousand years before fading away, eventually to be reincarnated as a modern version in 1896 in Athens. The Olympic flame, however, was not reintroduced until the 1928 Games in Amsterdam, and it wasn't until the 1936 Games in Berlin that the relay reappeared, devised by history professor Carl Diem. The flame was lit with a mirror at Olympia in Greece, just like in ancient times, and then carried by over 3,000 runners to Berlin. There was a disturbing undercurrent to the relay, because Hitler regarded ancient Greece as the Aryan forerunner of the Third Reich and his propaganda machine under Joseph Goebbels aimed to use the Olympics to prove that Germans were the master race. American athlete Jesse Owens destroyed that abhorrent notion by running faster and jumping farther than any German.

Since 1936, the torch relay has been part of every Olympics, with the flame sometimes traveling by air, boat, or even laser beam. Open flames are not welcome on airplanes, so a safe method of transportation had to be devised. The solution was essentially the same as the one Humphry Davy came up with some 150 years earlier to prevent the flame in miners' lamps from triggering explosions in coal mines. Enclosing the flame in a fine wire mesh cage allowed the flame's heat to be readily dissipated. In 1976, for the Montreal Olympics, the flame at Olympia was converted to a radio signal for transmission across the ocean via satellite. Since a flame can conduct electricity, it can be inserted into an open circuit and the current flowing through it can be converted to a radio signal that can be bounced off a satellite. At the other end, the radio signal can be used to activate a laser beam that lights a flame.

Leading up to the 2000 Olympics in Sydney, the torch took a three-minute underwater trip across Australia's Great Barrier Reef, and during the Beijing Olympics' torch relay, the flame traveled to the top of Mount Everest. Both of these required specially designed torches that had to supply both fuel and oxygen. Any combustion process requires a fuel that burns and an oxidizing agent that supports combustion. Since the 1972 Munich Games, the fuel has been either propylene or a mix of propane and butane, housed in a canister inside the torch. The latter combo produces less soot and burns brighter. But since the oxygen content of the atmosphere decreases with elevation, oxygen at the top of Mount Everest is in short supply. So a special torch was designed that used ammonium perchlorate as the oxidizing agent. When this compound decomposes, it yields water, hydrogen chloride, nitrogen, and oxygen. The fuel used was aluminum, which burns to yield aluminum oxide with an especially bright flame due to the glowing particles of aluminum oxide. This is actually the same chemistry that was used in the solid fuel boosters of the Space Shuttle where the propellant mixture consisted of 70 percent ammonium perchlorate and 16 percent aluminum, with the rest being made up of iron oxide that served as a catalyst and a polybutadiene-acrylic rubbery binder to give Ammonium Perchlorate Composite Propellant the consistency of a rubber eraser.

Not only is there an oxygen deficiency problem underwater, but the cooling effect of the water also has to be contended with. Although torch designs are generally somewhat secretive, the consensus is that the Sydney underwater torch used magnesium and aluminum as fuel with barium sulphate and barium nitrate as oxidizing agents. Magnesium burns with an extremely bright and hot flame, which is great underwater, but when the same fuel mixture was used to give a little extra flair to the torch used to light the cauldron at the Melbourne Olympics in 1956, the runner ended up with some nasty burns from the glowing bits of magnesium.

The London 2012 torch had no underwater requirement, but it did have to stand up to the British rain and gusty winds and had to be light enough to be carried for 400 meters without burning the bearer's hand. These problems were solved by boring 8,000 tiny holes into the triangular aluminum alloy framework using a super-sophisticated laser. The 8,000 holes represent the 8,000 people who took part in the torch relay over 8,000 miles. The triangular shape was chosen to symbolize the three times that London has held the Games, as well as the "faster, higher, stronger" motto. Gold, of course, is the most coveted metal at the Olympics and the designers decided to incorporate its color into the torch. But the color is not due to actual gold; it is the result of depositing a fine layer of titanium nitride on the aluminum alloy using cutting edge physical vapor deposition technology. I think Edward Barber and Jay Osgerby, the designers of the London Olympic torch, deserve a gold medal!

Chemical Demonstrations Can Get Mighty Hot

You know the old expression that a picture is worth a thousand words? Well, when it comes to chemical demonstrations, a video is worth a thousand pictures, and a live performance is worth a thousand videos! OK, maybe not a thousand, but several. At one time, demonstrations were an integral part of chemistry lectures but these days they tend to be rare. Good demos take time to prepare, take time to set up, and often require a cleanup. Videos, on the other hand, are easy to show and many excellent ones are available. But the same way that a movie can never quite capture the thrill of a live stage performance, a video of a combustion reaction just doesn't have the same impact as a professor performing it live in front of a class.

One of the features of a live demo that students find especially appealing is the chance that something may go awry. And with combustion-type reactions, that can certainly happen. I know. I once came pretty close to a nasty accident with the classic "burning money" demonstration in which a bill is immersed in a mixture of water and alcohol and is set ablaze. The flames can be clearly seen even in a large lecture hall, and to the students' amazement, when they are extinguished, the bill is totally undamaged, albeit a little wet. How does this happen? It's a case of the alcohol burning and the water preventing the bill from catching fire. Once when I was performing this, I accidentally tipped over the beaker holding the alcohol-water mixture, which then caught fire. Much merriment ensued when the students saw my flaming hands. Luckily, nothing else on the desk caught fire and my hands were protected by the water the same way the bill was. But now when I perform this demo, I always make sure the beaker is securely clamped to a stand.

My little accident was nothing compared with what was likely the worst chemical demonstration accident ever. That occurred in 1957, at Indiana University in Bloomington, when a professor demonstrated the effect of liquid oxygen on the combustion of aluminum in front of a group of high school students, hoping to give them a memorable chemical experience. It turned out to be memorable, indeed, but not in the fashion he envisaged.

Oxygen itself does not burn, but it does support combustion. Aluminum, on the other hand, burns well, especially when powdered to provide a large surface area for potential contact with oxygen. Dowsing the powdered aluminum with liquid oxygen creates ideal conditions for combustion. Most of the liquid oxygen quickly evaporates, but some gets trapped in the crevices of the tiny aluminum particles. The classic demonstration of the "liquid oxygen effect" involves using a candle to set fire to a sample of powdered aluminum in a metal crucible after dowsing it with liquid oxygen. The usual result is a bright flare that shoots straight up in a spectacular but harmless fashion.

Up to 1957, this experiment was a standard one, frequently performed at universities around the world without any problem. But on that fateful day in Indiana, instead of just burning brightly, the mixture detonated, hurling fragments of the iron crucible and the stone tabletop on which it had been sitting throughout the auditorium, causing injuries that ranged from the loss of an eye to severe lacerations and even crushed bones. It turns out that under the right, or in this case, wrong conditions, the reaction inside the mound of aluminum was so rapid and produced so much heat that the air trapped inside the crucible expanded with tremendous speed and produced a shockwave. And that is what we call an explosion! Although with great care the liquid oxygen-aluminum reaction can be performed safely, this is one demonstration I wouldn't touch with a 10-foot, or even a 100-foot, pole. But it is perfectly safe on a video.

There's another demonstration that I once performed but do not care to repeat. The "thermite reaction" is just too dangerous. But it is a doozy. Once more, powdered aluminum is involved, this time mixed with finely ground iron oxide, familiar to us as "rust." When this mixture is ignited, the oxygen is transferred from the iron oxide to the aluminum, yielding aluminum oxide and metallic iron. Ignition is not easy, but can be achieved with a strip of magnesium. Magnesium lights readily and the heat generated ignites the thermite mixture, which then proceeds to produce a tremendous amount of heat with the temperature of the products reaching some 2,200°C, or 3,992°F! Glowing bits of iron spew out in a shower of sparks, impressing everyone around, especially those who have an encounter of the first kind.

Why would this reaction be worthy of a demonstration? There is, of course, the gee-whiz effect. But there's more. The thermite reaction has some practical applications. Soon after it was patented by German chemist Hans Goldschmidt in 1895, this reaction was put to use in welding, especially when it came to joining railroad tracks. And humans, being humans, soon found a military application in incendiary bombs. These were used with great efficiency both by the Germans and the Allies during World War II. Thermite hand grenades were also developed for the destruction of captured military equipment as well as for the emergency destruction of sensitive equipment in the face of an impending risk of being captured by the enemy. Exploding a thermite grenade in the barrel of a cannon will ensure that the weapon is permanently disabled. And exploding one inside a helicopter ensures its destruction. U.S. Navy SEALs likely used thermite grenades to destroy the secretly developed stealth helicopter that crashed during the assault on Bin Laden's compound in Pakistan.

Conspiracy theorists also maintain that the thermite reaction was used by the "master architects" of 9/11 (their term for the U.S. government) in order to bring down the Twin Towers from the inside because, according to these sages, burning jet fuel is not hot enough to melt structural steel. What evidence do they provide? A video that shows an automobile engine being melted by a thermite reaction. Such absurd conspiracy theories deserve to crash and burn.

The Many Faces of Neoprene

Diving suits, gaskets, hoses, life rafts, iPad covers, and giant balloons destined for Macy's Thanksgiving Day parade. What do they have in common? All are made of neoprene! Not only does this synthetic rubber have myriad uses, but it also holds a place of honor in history for having ushered in the age of modern plastics.

Neoprene was first synthesized in 1930 by DuPont chemist Wallace Carothers, who would later achieve worldwide fame as the inventor of nylon. It was born out of the need to find a substitute for natural rubber, an item that by the first decade of the twentieth century was becoming nearly as indispensable as coal or iron. The automobile industry and the military were particularly reliant on rubber, but the natural substance, an exudate of the rubber tree, was often in short supply and reacted too readily with oxygen, meaning that it did not age well.

Attempts to synthesize rubber trace back to 1860, when English chemist Charles H. Greville Williams successfully degraded natural rubber to its basic building block, a simple compound called isoprene. This germinated an idea: why not take some isoprene and "reverse engineer" it into rubber? It turns out that isoprene can be readily isolated from the mixture of compounds that form when petroleum is heated, a process known as cracking. But converting isoprene into rubber proved to be a formidable task. Failure followed failure until World War I rolled around and a British blockade forced the German hand.

As early as 1910, German chemists had experimented with producing rubber from methyl isoprene, a close relative of isoprene. Although its properties were less than ideal, Germany's desperate need for rubber to sustain the war effort pressed methyl rubber into service. The search for a better product continued until the 1930s, when Dupont chemists, led by Wallace Carothers, came up with neoprene, the first truly useful synthetic rubber. But that discovery would not have come about were it not for the pioneering work of Father Julius Nieuwland, a Roman Catholic priest who had taken up a post as a professor of chemistry and botany at Notre Dame University.

As a graduate student, Nieuwland had become fascinated with acetylene, a gas he would end up investigating for the rest of his career. At one point, he reacted acetylene with a copper catalyst to produce a yellowish oil, identified as divinyl acetylene. Much to his surprise, when left to stand, the oil thickened into a jelly and then into a hard resin. The reaction wasn't readily reproducible and Father Nieuwland began to work at fine-tuning the process. In 1923, he hit pay dirt by reacting his divinyl acetylene with sulphur dichloride to produce a substance with elastic properties. To Nieuwland, this was of great theoretical interest and merited a presentation at the American Chemical Society's meeting in Rochester, New York. As chance would have it, Dr. Elmer Bolton, who headed a DuPont project to make synthetic rubber, happened to attend Nieuwland's talk and immediately became interested in using divinyl acetylene as a starting material. The company struck an agreement with Father Nieuwland for the use of his catalyst and DuPont chemists began to produce a variety of rubbery materials, none of which were satisfactory.

Then, in 1930, the brilliant Carothers, years before his epic discovery of nylon, was asked to look into the problem. He suggested trying starting materials closely related to divinyl acetylene. The breakthrough came with Carothers's idea of using hydrogen chloride to try to link molecules of monovinyl acetylene together. Arnold Collins, one of Carothers's assistants, worked on this reaction and found that monovinyl acetylene reacted with hydrogen chloride to produce a new liquid that was christened "chloroprene." To Collins's glee, upon standing, the liquid solidified into a rubbery substance that bounced when dropped on the lab bench. As Carothers would explain, the small molecules of chloroprene had linked together to form polychloroprene. DuPont originally named its new rubber "DuPrene" but later renamed it "neoprene," a more generic term since the company only made the raw material, not any finished product. The synthesis of neoprene was destined to become a milestone in the development of polymers.

Neoprene had greater tensile strength than natural rubber and better resistance to oxygen, chemicals, and abrasion. By the time the U.S. entered World War II, neoprene was being produced on a large scale at DuPont's plant in Louisville, Kentucky, eventually resulting in the city being nicknamed "Rubbertown." During the war, Japanese forces managed to cut off the U.S.'s natural rubber supplies from Malaysia, but thanks to the availability of neoprene, the effect of the embargo was greatly reduced.

Uses of neoprene were not restricted to the military effort. Father Nieuwland had a pair of heels made for his shoes as he traveled through Europe in 1934. When the soles wore out, he had the heels transferred to another pair. Since that time, neoprene has found numerous applications. It is one of the components of space suits, and its ability to be "foamed" has made wet suits possible. Foaming allows for bubbles of nitrogen gas to be incorporated into the material for insulation purposes. The softness of foamed neoprene makes it an ideal material for the protection of laptops, cell phones, and iPads.

Due to his vows of piety, Father Nieuwland never accepted any royalties for his invention. For DuPont, however, neoprene was a great success in many ways. The product sold very well, but perhaps even more importantly, it stimulated Carothers to delve into the chemistry of neoprene. It was clear that the novel material was a polymer, formed by linking the small molecules of chloroprene together. A series of classic papers by Carothers in the 1930s put polymer chemistry on a firm footing and resulted in his invention of polyesters and then, famously, nylon.

Father Nieuwland is also known for his discovery of the chemical weapon Lewisite during his acetylene research. Nicknamed "dew of death" on account of its terrible blistering effect, Lewisite was produced by the U.S. but never used in warfare. When the compound he had discovered was turned into a weapon, Nieuwland nearly gave up chemical research entirely. Good thing he didn't, or today we might not have have neoprene or its descendants.

From Twitching Worms to Non-Browning Apples

The tiny worm's twitch was hardly noticeable, but with that slight shudder, science took a giant leap! A leap big enough to lead to a Nobel Prize that would pave the way to apples that will not brown, onions that will not make you cry, cottonseeds that you can eat, and diseases that you can treat.

The 2006 Nobel Prize in Physiology and Medicine was awarded to Professors Andrew Fire of Stanford University and Craig Mello of the University of Massachusetts for their discovery of RNA interference and its role in gene silencing. Genes are those segments of the "master molecule of life," DNA, that speak, but not with words. Their language is expressed in molecules, specifically ones known as messenger RNA or mRNA. The message they carry is the set of instructions for the construction of proteins.

Life is all about proteins. Not only are these molecules the building blocks of our tissues, they make up the antibodies that protect us from disease, the receptors that allow cells to communicate with each other, and the enzymes that catalyze virtually every reaction that goes on in our bodies. But how do cells know which proteins to make? That's where the 30,000 or so genes dispersed along the strands of DNA come in. Each gene holds the instructions for making a particular protein, but the problem is that proteins are synthesized not in the nucleus but in the cytoplasm of a cell. How, then, does the message get from the DNA in the nucleus to the protein-making machinery in the cytoplasm? By means of the appropriately named messenger RNA. If this process is interfered with, the protein the gene codes for doesn't get made, and the gene is effectively silenced.

Now back to our little nematode worms. Some of these creatures make twitching movements because they lack a protein needed for proper muscle function as a result of having a non-functional gene. Fire and Mello's breakthrough discovery involved making normal worms twitch by silencing the appropriate gene through injection of a special type of RNA (double-stranded RNA). It turns out that if this tailor-made RNA matches the genetic code of a specific messenger RNA, it will inactivate it, thereby essentially silencing the gene that triggered the formation of that particular messenger RNA.

Subsequent research showed that this RNA interference machinery can be activated in yet another fashion, without the introduction of any double-stranded RNA from the outside. Sometimes, for proper functioning of our bodies, the synthesis of certain proteins needs to be suppressed; some genes have to be silenced. Cells accomplish this through making double-stranded RNA via an intermediary known as micro-RNA, which in turn is synthesized on instructions encoded in the cells' DNA. In other words, DNA contains genes that can silence other genes through RNA interference.

With the difficult theoretical stuff out of the way, let's get down to some practicalities. The world has no need to remedy muscular problems in worms, but how about producing apples that do not turn brown? At first this may seem like a frivolous application of RNA interference, but that is not necessarily the case. A Canadian biotechnology company, Okanagan Specialty Fruits (OSF), has developed a non-browning apple by silencing a gene that codes for an enzyme known as polyphenol oxidase (PPO).

When an apple's cells are ruptured by bruising, slicing, or biting, PPO and oxygen from the air combine with naturally occurring phenols in the apple to trigger a chemical reaction that forms melanin, a brown substance that is thought to protect the apple from attack by microbes. But the brown discoloration is unappetizing and often results in apples being discarded. The traditional way of preventing such browning is with lemon or pineapple juice, the acidity of which inactivates polyphenoloxidase. Commercially packaged apple slices are usually dipped in an antioxidant solution of calcium ascorbate. Genetically modified apples that do not brown would not require either treatment. And sliced apples that do not brown would avoid the yuck factor and make for a healthy addition to children's lunches. Any method that allows for greater apple consumption is attractive.

The exact fashion in which the "Arctic apple," as it will be known, is genetically modified is proprietary information, but it is accomplished through RNA interference. Here is a possible way. Some apples are naturally very low in polyphenoloxidase because they express a gene that codes for the double-stranded RNA that in turn silences the PPO gene. Through standard genetic modification methods, this silencing gene can be copied and inserted into the DNA of other apples with the result that PPO production will be silenced and the apples will not turn brown.

Not everyone is thrilled by the possibility of genetically altering apples in this fashion. Organic growers worry that pollen from the modified apple trees will spread to their orchard, potentially causing them to lose their organic status. Okanagan Specialty Fruits argues that apple pollen does not blow around easily and the chance of it spreading to a neighboring orchard is slim. Some critics, particularly anti-GMO activists, have suggested that silencing the PPO gene may have unintended negative consequences, but there is no evidence for this. That comes as no surprise because there are no novel proteins being formed. Field trials have shown that the modified apples are like all other apples except that they do not turn brown.

Using RNA interference technology, the lachrymatory factor synthase gene in onions can be silenced so that the nutritional qualities of this vegetable can be enjoyed without weeping. And how about cottonseed? The world produces some 44 million tons of high-protein seed every year that cannot be eaten because it contains the poisonous compound gossypol. Using RNA interference, the gossypol-producing gene can be silenced and enough protein to meet the daily requirements of half a billion people can be produced. But perhaps the most alluring potential of RNA interference lies in tackling genetic diseases. There have already been some preliminary successes, albeit only in mice, with silencing genes that code for toxic proteins such as the ones found to be present in Huntington's disease, as well as in silencing genes that cause high cholesterol levels.

In the meantime, the Canadian Food Inspection Agency is considering an application to market the "Arctic apple," and a ruling is expected in 2014. Whether it makes it to market or not, there is no doubt that the journey from twitching worms to non-browning apples has been a fascinating one! Let's hope we won't end up with worms in the apple by silencing the polyphenoloxidase gene.

A Squeeze on Orange Juice Production

We all have habits. One of mine is to start every day with a glass of orange juice. Freshly squeezed would be ideal, but frankly, too much of a bother. So I go for the one that says "100 percent pure and natural, nothing added, nothing taken away, just oranges." My preference has nothing to do with the "pure" or "natural," it has to do with taste. But I certainly took note when I came across a report that a class-action lawsuit had been launched in the U.S. against my favorite juice, Tropicana. What horrific toxin had been uncovered, I wondered? Actually, none.

The lawsuit asserts that the label claim "100 percent pure and natural" is false because the juice is in fact heavily processed and its flavor is scientifically manipulated. That being the case, the plaintiffs allege that they would not have purchased the juice, or paid more than they otherwise would have been willing to pay, had the juice been properly labeled. While this lawsuit has a frivolous flavor, it does bring the discussion of the term "natural" into the public arena. That's a good thing, because this expression is widely bandied about in a meaningless fashion in order to cater to the common fallacy that "natural" is always superior to "processed," "artificial," or "synthetic."

There is no question that the image projected by Tropicana is that oranges are squeezed, then their juice is poured into a carton, which is then speedily delivered to a store near you. But that is not what happens. The truth is that if you want fresh, great-tasting orange juice, there is only one way to get it: squeeze the oranges yourself and drink the juice as quickly as possible. Why quickly? Because the squeezing process unleashes a cavalcade of chemical reactions, most of which are not favorable. Enzymes released from the ruptured cells catalyze reactions between oxygen in the air and some of the hundreds of compounds that occur naturally in orange juice. The oxidized products result in off-flavors. If orange juice from a carton is to taste fresh, these reactions have to be curtailed, which is why Tropicana has been working since the 1950s on various technologies to maintain as much of the original flavor as possible. And on ways to ensure that the juice contains no harmful bacteria.

Like virtually all foods, oranges can harbor bacteria. Pasteurization, basically heating the juice, kills bacteria, but it also kills taste. So, unpasteurized juice may appeal to the taste buds, but sadly, there have been well-documented cases of poisoning from such juice, with perhaps the most famous one being in an Orlando theme park where more than sixty people came down with salmonellosis. How, then, do you get rid of bacteria and preserve the taste? Back in 1952, Anthony Rossi, who founded the company that would eventually become Tropicana, introduced a flash pasteurization process that managed to preserve the flavor by raising the temperature of the juice for only a very short time.

Pasteurization is critical since oranges are seasonal and juice sometimes has to be stored as long as a year to satisfy consumers who like to start each day with this beverage. And why shouldn't they? It is a great source of potassium and provides a large array of antioxidants. But of course there's also a large dose of sugar, comparable to soft drinks, so we don't want to be guzzling OJ by the gallon.

Orange juice from frozen concentrate was already available in the 1950s, but its taste was far removed from that of fresh oranges because the heat treatment needed to concentrate the juice resulted in the breakdown of flavor compounds as well as in the formation of novel compounds with undesirable characteristics. Development of Rossi's flash pasteurization was followed by the introduction of deaeration, a process by which a vacuum system removes much of the air from the freshly squeezed juice to prevent oxidation reactions. Unfortunately, this also removes some of the natural orange flavors. But there was a scientific solution to this as well. The aromas were collected, blended with orange oil extracted from the peel, and then added back to the juice.

The volatiles and orange oil make for a very complex mix of hundreds of compounds that can be separated from each other by techniques such as distillation and chromatography. A chemist can then assemble these into flavor packs with varying tastes. One of the major compounds responsible for fresh orange taste is ethyl butyrate, so flavor packs that contain higher levels of this chemical are especially desirable. Since oranges differ in taste depending on their variety and the time of the year they are picked, the use of flavor packs allows for uniformity of taste year round. Basically, then, flash pasteurization, deaeration, and the blending of orange compounds into flavor packs allow for a juice that tastes as close as possible to freshly squeezed. But it isn't exactly "100 percent pure and natural," although that to some extent depends on the interpretation of these ambiguous terms.

Since all components of the juice do come from oranges, the juice can be called "natural." In common usage, however, "natural" has come to imply unprocessed, so in that way Tropicana is not natural. That, though, has nothing to do with the quality of the juice, which is excellent, as is evidenced by its popularity around the world. The real problem is the golden aura with which the term "natural" has been anointed. If oranges picked in Florida were just squeezed and the juice packaged and shipped up north, you would be drinking truly "natural" orange juice, but chances are high that you would not enjoy the natural bacteria and natural oxidation products.

Technically, the plaintiffs in the lawsuit against Tropicana are correct, and perhaps the labeling should be changed. Indeed, why not emphasize the scientific ingenuity in overcoming the natural degradation of orange juice? Here's my suggestion: "The best-tasting orange juice that nature and modern science can deliver." Because it really is.

Sleeping Gas — It's a Dream!

Far too often, we've seen news reports of police confronting holed-up criminals or terrorists. Why don't they use sleeping gas to put an end to the situation? Well, they probably would, if there were such a thing as a true sleeping gas. While there is plenty of doubt about the existence of a sleeping gas, there is no doubt that there is a widely held belief that such a chemical exists.

When a pharmaceutical tycoon's villa was robbed in Sardinia in 2011, speculation was that the burglars used some sort of "sleeping gas." How else would the occupants not have noticed that the criminals had rummaged through the house, stealing loads of cash and jewelry after taking windows off their hinges? Stories also circulate about thieves in France and Spain spraying gas into the air vents of camper vans before robbing the sleeping occupants.

The concept of a sleeping gas being used by criminals is not new. Back in a 1952 television episode, the Cisco Kid, a popular western hero, confronted a gang of outlaws who used sleeping gas to carry out a series of robberies. And in the James Bond classic, *Goldfinger*, the villain's accomplice, Pussy Galore (amazing that they got away with that name), was told that her airplane would be equipped with sleeping gas to knock out the soldiers who were guarding Fort Knox. In fact, the use of sleeping gas in Europe is likely to be just as fictional as the Cisco Kid or *Goldfinger* stories. There simply is no gas that can be released in a confined area that will reliably knock people senseless then have them wake up with no significant aftereffects.

Of course, there are anesthetic gases that can induce sleep, although "gas" is a bit of a misnomer here. The most widely used inhalation anesthetics such as isoflurane, desflurane, or sevoflurane, are liquids at room temperature. They are, however easily volatilized. Nitrous oxide is a true gas, but it does not produce surgical-level "sleep" unless combined with another agent. Beginning with ethyl ether in the mid-nineteenth century, a number of inhalation anesthetics including chloroform, cyclopropane, enflurane, ethylene, halothane, trichloroethylene, and vinyl ether have found use, but all require careful administration through a mask and continuous monitoring of the patient's level of consciousness. None of these agents will induce sleep just by being released in a room.

The only documented attempt to use a "sleeping gas" takes us back to 2002 and the famous hostage-taking episode in a Moscow theater. A group of Chechen terrorists stormed the venue during a performance, taking some 850 hostages and demanding that Russian forces withdraw from Chechnya. After days of fruitless negotiations, authorities decided to pump a gaseous incapacitating agent into the theater. The hope was that everyone inside would be subdued before special forces burst in to capture the terrorists and liberate the hostages. It didn't turn out that way. Most of the terrorists either died from the gas or were killed during the assault. Tragically, 129 hostages also met the same fate. All but a few died as a direct result of inhaling the gas that was only supposed to put them to sleep.

The exact nature of the gas remains a mystery. Russian authorities have maintained secrecy, but consensus is that the most likely candidate for the gas is an aerosol consisting of 3-methylfentanyl dissolved in halothane. There is some interesting history here that begins with the first synthetic opiate, meperidine, widely known as Demerol.

Natural opiates, such as morphine and codeine, are found in the milky latex that exudes from the bulb of the opium poppy. These compounds are long-standing effective pain killers, but they come with some baggage. Impaired respiration is a side effect and there is always the problem of addiction. As synthetic organic chemistry matured, many pharmacological chemists focused on creating new compounds with slight variations on the molecular structure of morphine. The hope was that these analogues would in a sense separate Dr. Jekyll from Mr. Hyde, retaining the favorable properties of morphine while eliminating the nasty ones.

Besides dulling pain, opiates can also reduce muscle spasms. Meperidine was developed in 1932 as an antispasmodic alternative to morphine, with hopes that it would be safer and less addictive. While this hope did not materialize, meperidine did turn out to have useful pain-killing properties and took on a role alongside morphine. But the search continued for the Holy Grail, an opiate analog devoid of side effects. In 1960, Paul Janssen synthesized fentanyl, based on the meperidine model. This compound was far more potent than morphine or meperidine, but on the downside, its painkilling duration was brief. Fentanyl's first use was as an intravenous anesthetic, but by the 1990s, the fentanyl patch as well as a "lollipop" made with fentanyl citrate brought effective pain relief to many. Because it is more soluble in fatty tissue than morphine, fentanyl enters the central nervous system more quickly and affords faster action. Unfortunately, this also makes the drug prone for abuse. Users can become addicted quickly, and fatal overdoses due to the powerful respiratory depression effect are not uncommon.

Fentanyl is a controlled substance, but leave it to clandestine chemists to cook up a batch and put it out on the street. Even more scary is their ability to come up with analogs that are more potent and more dangerous. In the 1970s, 3-methylfentanyl began to appear, often under the name "China White," promoted as an "improved" version of heroin. That improved version cut a swath of death among addicts.

Like all opiates, 3-methylfentanyl can put people to sleep. Often permanently. This is the drug that many believe was used by the Russians in the attempted rescue of the hostages. The suspicion arises from the instructions given by the authorities to physicians who were administering to the hostages after the "rescue." They were told to inject naloxone, a well-known blocker of opiate activity. Whether this was useful remains a mystery because details of the whole operation were never released. It is clear, though, that whatever was used was not a "sleeping gas." The existence of such a substance — used either by criminals or against them — remains mired in mythology.

The Electrifying Lectures of Sir Humphry Davy

The long queue snaked along Albemarle Street in London. Were the elite of British society waiting in line for an opera performance? No. A royal gala? No. A star-studded movie premiere? Hardly, considering this was June 20, 1801. Believe it or not, the attraction was a public lecture on chemistry!

The rich and famous had come to the Royal Institution to hear young Humphry Davy, whose scientific "performances" had become the talk of the town. And they would not be disappointed. As his lecture on "Pneumatic Chemistry" drew to a close, Davy administered "laughing gas" to a number of volunteers, much to the merriment of the audience. The epic event is commemorated in a classic caricature by satirist James Gillray, in which an impish-looking Davy looks on as the nitrous oxide he has just prepared makes a rather explosive exit after being inhaled by an unfortunate subject.

Although Gillray pokes fun at Davy's antics, the young chemist's lectures had a serious impact on the forward march of science. Not only did these presentations mark the beginning of a commitment by the Royal Institution to popularize science, but Davy's drive to captivate the growing audiences with novel chemical phenomena also led him down a research path that resulted in a number of groundbreaking discoveries. By the time of his death in 1829, Sir Humphry Davy had established a legacy that ranged from the discovery of a number of elements to saving miners' lives with the "Davy safety lamp." All this despite the lack of a formal education.

Growing up in Penzance, young Davy showed a scientific bent, but his surgeon grandfather decided that the boy should be apprenticed to a physician. Since at the time doctors also served as apothecaries, formulating medications gave the lad an opportunity to hone his chemical skills. These skills caught the attention of Dr. Thomas Beddoes, a physician who had some strange ideas about treating tuberculosis.

Beddoes had observed that butchers rarely suffered from the disease and came to the conclusion that this must have something to do with being around cattle. So he set up a clinic where patients shared rooms with cows so that they could benefit from the bovines' exhaled breath and potent flatus. The doctor was apparently satisfied with the results, because he decided to establish a "Pneumatic Institute" where patients would be treated with a variety of gases. But he needed someone with chemical know-how to generate them. Thus the job offer to Davy, who had already discovered that he would rather deal with chemicals than with patients. A disappointed grandfather promptly wrote Humphry out of his will.

At the Pneumatic Institute in Bristol, Davy focused his attention on nitrous oxide, a gas first synthesized in 1772 by Joseph Priestley, who made it by heating ammonium nitrate. Davy dubbed it "laughing gas" after watching the amusing effects it had on visitors to the Institute. He tried it himself, recording how "it made me dance about the laboratory as a madman, and has kept my spirits in a glow ever since." Davy even noted, "as nitrous oxide appears capable of destroying physical pain, it may probably be used with advantage during surgical operations in which no great effusion of blood takes place." Curiously, this idea lay dormant for another forty years until Horace Wells, a Boston dentist, resuscitated it after noting that a friend had become insensitive to pain after inhaling the gas for recreational purposes. Indeed, it was Davy's experiments at Beddoes's facility, and later his classic lecture at the Royal Institution, that triggered the public's fascination with laughing gas and made its inhalation a staple at traveling medicine shows and carnivals.

The Royal Institution was founded in 1799 by fifty-eight well-to-do gentlemen from various walks of life who shared a passion for science and a conviction that affording the public access to scientific knowledge was the key to eventually applying that knowledge for practical purposes. The stated goal was "diffusing knowledge, and facilitating the general introduction of useful mechanical inventions and teaching by courses of philosophical lectures and experiments, the application of science to the common purposes of life." Thomas Garnett was hired as the first Professor of Chemistry at the Institution, and in 1800, he delivered the first public lecture.

It seems the Institution's managers were not totally pleased with Garnett's efforts and looked for a way to appeal to the public in a broader fashion. The flamboyant young man who entertained the clientele at Beddoes's Pneumatic Institute with his gases might be just the one to attract wealthy patrons with his exuberance! Furthermore, his good looks seemed sure to appeal to the gentle sex. And, as the Institution's managers were to learn, Davy had more tricks up his sleeve than just amusing people with laughing gas.

The chemist had been electrified by the work of Alessandro Volta, who had devised a battery consisting of alternating zinc and copper plates separated by paper soaked in a brine solution. If a chemical reaction could generate electricity, Davy postulated, perhaps electricity could also generate chemical reactions.

He had already carried out some preliminary experiments along these lines, which the Institution's managers thought would make for great demos to entice the public. An offer was made to Davy, and the rest, as they say, is history.

Not only did the colorful Davy deliver spectacular lectures, he also established the Royal Institution as a hotbed for scientific research. It was in the Institution's laboratory that he himself first isolated potassium by passing an electric current through molten potash, and then sodium by the action of a current on molten sodium hydroxide. He later followed with the discoveries of magnesium, calcium, strontium, and barium.

Davy's lectures and demonstrations were so brilliantly presented that they became fashionable social events. Traffic jams created by carriages bringing the crowds to the Royal Institution lectures resulted in Albemarle Street being made into the first one-way street in London.

It is fair to say that anyone who aims to improve the public understanding of science should pay homage to the pioneering efforts of Sir Humphry Davy. Today, the need to demystify science is possibly greater than ever and it is enlightening to reflect upon the serious work that began with a whiff of laughing gas.

The Ups and Downs of Antibiotics

Talk about a double-edged sword. Antibiotics, when properly used, save lives. But, like all drugs, antibiotics do come with some baggage. The major issue is drug resistance. Bacteria, even of the same species, are not totally alike. Some have a greater natural resistance to antibiotics than others. This should come as no surprise; the phenomenon of biological individuality is evident in humans. Not everyone exposed to the same virus or bacterium gets sick. As far as bacteria go, the ones that survive an antibiotic onslaught will go on to reproduce and pass their genes on to their offspring, conferring resistance to the antibiotic in question. Next time that antibiotic is used, it will be less effective because the resistant population is now much greater. Basically then, every time an antibiotic is used, the probability that it will be effective the next time around is reduced.

When someone has a serious bacterial infection, physicians should of course reach for the prescription pad. But what about the farmer who treats his animals with low doses of antibiotics in an effort to protect them from disease and make them gain weight more rapidly? Although generally these are not the same antibiotics used to treat illness in people, there is evidence that bacteria can transfer genes to each other, and that some of the bacterial resistance we are now seeing in people may be the result of incorporating antibiotics into animal feed.

The U.S. Food and Drug Administration recently put out an advisory about the overuse of antibiotics in animals and is asking for voluntary compliance in limiting the use of these important drugs when treating disease in animals and eliminating their use for "production" purposes. Food producers, drug companies, and veterinarians are being asked to cooperate in making the changes in the way antibiotics are used in animals, but the FDA has stated that if voluntary action is not forthcoming, legal restrictions such as those already in place in Europe will be instituted. That, however, is likely to be a messy business with back-and-forth court actions. Hopefully, with the realization of the serious stakes involved, producers and antibiotic manufacturers will comply with the request to alter the way that antibiotics are used in animals both in the U.S. and in Canada.

While it has long been known that low-dose antibiotics cause weight gain in animals, the mechanism by which they do this has been a mystery. Researchers are now beginning to zero in on the effect, and their work may even shed light on the human obesity epidemic. Animals, like humans, have numerous bacterial species living in their gut. Believe it or not, there are more bacterial cells in our body than human cells. Of course, bacterial cells are much smaller than human cells. But their effect on our health may not be small.

Some varieties of bacteria are more likely to cause the body's immune system to swing into action, but usually different bacteria keep each other's multiplication rate in check by competing for the same food supply. But if the bacterial balance is upset because an antibiotic reduces the numbers of one species more than others, an inflammatory response can occur. Such a response is linked with making our cells less sensitive to insulin. Insulin resistance means that glucose is less likely to be taken up by cells, and since it is the cell's main source of energy, they crave an increased intake. This translates to a boost in appetite as the body strives to meet cellular needs.

What all of this suggests is that some species of bugs in our intestine may contribute to weight gain more than others, and that these may become more prevalent when competitors are reduced by antibiotics. Of course, other factors may also play a role in altering the bacterial flora. The chlorination of drinking water as well as improved sanitation may influence both the type and the number of microbes that reside in our gut. Who knows, perhaps all that emphasis on getting rid of germs may be affecting our waistlines.

Is there any actual evidence for this postulated link between changes in gut bacteria and obesity? In one word, yes. When Martin Blaser, a microbiologist at New York University, fed infant mice doses of penicillin comparable to those given farm animals, he found that after thirty weeks, these mice had put on

10 to 15 percent more weight than those not treated with the antibiotic. Furthermore, the mice that had been treated had a different microbial flora in their gut, with *Lactobacillus*, one of the "good" bacteria, having been significantly decreased. When gut bacteria from these mice were introduced into mice that had been bred in a totally sterile environment, and were therefore germ-free, they put on more weight than mice with the regular complement of microbes in their gut.

But mice are not small humans. So, what about people? Danish researchers who followed the development of some 28,000 babies found that exposure to antibiotics within the first six months of life increased the chance of the children being overweight by the time they reached the age of seven. Conceivably, the effect of antibiotics on weight gain may even be passed down to offspring. Babies acquire bacteria as they pass through the birth canal, and if the mother had been treated with antibiotics as an infant, her weight gain–prone bacterial flora may be passed on to the baby, who then is predisposed to obesity.

While overconsumption of food is the crux of the obesity problem, one of the reasons why we eat too much, as we have now seen, may be due to the changes that have occurred in the microbial population of our gut. A study of ancient feces from caves, as well as from the intestinal tract of mummies, has revealed a microbial makeup that is quite different from that found in our guts today. Those ancient microbial populations are more similar to the ones found in rural African children, and some primates than in the intestines of North Americans, who are more likely to have been exposed to chlorinated water, antimicrobial cleaning agents, and antibiotics. Maybe a partial answer to obesity is to repopulate our intestines with the bacteria found in ancient poop. Can a diet that aims to woo with poo be far off?

The Saga of the Flaming Rocks

Lyn Hiner was happily munching on an orange in her kitchen when her pants spontaneously burst into flame, quickly filling the house with an acrid white smoke. As she struggled to rid herself of the burning garment, a couple of rocks tumbled out of a pocket. They were aflame! The ordinary-looking stones, clearly the cause of what would turn out to be second- and third-degree burns, had been collected earlier in the day on a California beach. They were the makings of a real chemical mystery.

Time for a flashback to World War II. A couple of German farmers walking through a cornfield noted some strange-looking pieces of plastic strewn about and pocketed a couple. Before long they were shedding their flaming trousers, victims of some British razzle-dazzle. Razzles, as they were commonly called, were a type of incendiary device designed to set fire to cornfields with hopes of crippling the German economy. They consisted of a layer of phosphorus sandwiched between two layers of perforated nitrocellulose, a notoriously flammable plastic. Dry phosphorus ignites upon exposure to the air, but is remarkably stable when moist, so razzles were kept immersed in water until they were dropped from airplanes. Once they hit the ground, they would dry out, ignite, and hopefully set fields ablaze. As it turned out, the razzle didn't really dazzle. A couple of flaming German farmers aside, the idea of setting fire to cornfields fizzled.

This was by no means the first time that phosphorus had been used in warfare. By World War I, an efficient method for producing phosphorus from calcium phosphate ore had been developed, allowing both the Allies and the Germans to use the element to produce smoke screens. When phosphorous ignites, it forms phosphorus pentoxide, which appears as a dense white smoke capable of obscuring an army from the enemy. But the most useful property of phosphorus in warfare is its ability to set things on fire, such as Zeppelins that were used by the Germans to bomb Britain. British fighter planes equipped with phosphorus-based incendiary shells brought down a number of Zeppelins, producing a dazzling display as the hydrogen that held them aloft ignited.

World War II revealed the truly horrific nature of phosphorus. Armies barraged each other with phosphorus-filled grenades and shells that burst to spew bits of burning phosphorus, causing terrible burns. But the ultimate was the phosphorus bomb, dropped from airplanes, capable of unleashing firestorms such as the world had not seen before. In Operation Gomorrah, thousands of firebombs dropped on Hamburg caused appalling injuries and ignited raging fires that practically wiped out the city. Quite ironic, given that it was in Hamburg back in 1669 that alchemist Hennig Brand discovered phosphorus by heating a vat of urine to dryness.

Earlier in the war, with the prospect of a German invasion looming, the British Government authorized the production of some six million Self-Igniting Phosphorus (SIP) grenades to be used by the population against enemy tanks. They were relatively simple devices consisting of phosphorus, benzene, water, and a piece of rubber sealed in a bottle. The rubber dissolved in the benzene to form a sticky mixture that would ignite when smashed against a hard surface. To boot, the burning vulcanized rubber released the choking odor of sulphur dioxide. It's a good thing, though, that SIPs never had to be put to use because tests on British tanks showed them to be ineffective. They did, however, produce some spectacular firework displays when some bottles stored in streams for safety broke loose, floated away, and crashed into rocks.

Japan also made use of phosphorus bombs during the war, albeit in somewhat of a low-tech fashion. Fu-Gos were balloons bearing a phosphorous bomb equipped with an explosive device designed to detonate on impact. Some 10,000 of these were released with hopes that they would drift across the Pacific and land in the U.S., causing fires and panic. There were a few casualties and fires, but Fu-Go turned out to be mostly a no-go.

Today, many fighter and bomber planes are equipped with phosphorus flares that can be launched when there is danger of attack by heat-seeking missiles. The missile's guidance system beams in on the burning phosphorus, diverting the weapon away from the plane. On occasion, such flares have been accidentally dropped into the ocean and have washed ashore. In Norfolk, England, in 2012, the bomb disposal squad was called in to blow up several flares that had been discovered on a local beach. Which now brings us back to the baffling rocks found on that California beach.

By all accounts the rocks looked perfectly ordinary and in no way resembled flares or any type of phosphorus-containing military device. Yet the events in the unfortunate victim's kitchen smack of burning phosphorus. The white smoke, the acrid scent, the deep burns, the difficulty in extinguishing the flames all point toward phosphorus. So does the finding by forensic chemists of phosphate residues on the rocks. The phosphorus pentoxide that forms when phosphorus burns reacts with water to form phosphoric acid, a source of phosphate.

So here is the scenario as it now appears. The rocks collected on the beach had somehow been treated with phosphorus, which was protected from reacting with the air by some sort of coating. As the rocks rubbed against each other in the unlucky lady's pocket, the phosphorus was exposed, and presto, ignition and misery! As to how these mysterious rocks ended up on the beach, everyone seems to be at a loss. Camp Pendleton Marine Base is nearby, as are several U.S. Navy installations, but their spokespeople have been unable to offer any explanation.

It is interesting to note that the word "phosphorus" derives from the Greek "*phos*" for light and "phorus," to bring. This is especially appropriate in this instance, given that what we really need is some light to be brought to the bizarre incident of the flaming California rocks.

Popping Off About Gluten-Free Rice Krispies

The malt flavoring is gone! Celiac sufferers are no longer limited to listening to the snap, crackle, and pop of Rice Krispies! They can actually eat the cereal that has been music to the ears of legions since 1928 but has been verboten for anyone with a sensitivity to gluten, the mixture of proteins found in wheat, barley, and rye. Rice contains no gluten and is in general a staple for celiac sufferers. But malt flavoring, a standard ingredient in Rice Krispies, can harbor a trace of gluten, which is enough to cause misery.

So what makes puffed rice speak to its fans? Some neat technology! First, the rice grains are treated with steam to introduce moisture, which performs a dual function. As more heat is applied, moisture provides the pressure needed to expand the grain of rice. Simultaneously, the water acts as an internal lubricant, or plasticizer, allowing the molecules of starch to slide past each other to meet the needs of the increasing surface area of the grain. As heating continues, water is expelled from between the starch molecules, which then form bonds to each other, setting up a rigid network that traps pockets of air. At the same time, some of the sugar, the second ingredient in Rice Krispies, dissolves, and then forms a tough film as the water evaporates, further strengthening the walls that surround the air pockets.

It is these air pockets that differentiate silent cereals from the musical ones. Every orchestra needs a conductor, and in this one, the baton is wielded by the milk. As the cold liquid is absorbed by the cereal, it squeezes out the trapped air, which then fractures some of the thin walls that separate the pockets producing the resounding snap, crackle, and pop! It seems, though, that when Rice Krispies were introduced, Kellogg's was not satisfied with a cereal that entertained the ears; it also had to entertain the palate. And that's when malt flavoring made its entry.

Remember going down to the malt shop with a date for a malt? If you do, you date yourself but will probably recall the soda jerk spooning some powdered "malted milk" into a glass, adding water, stirring, and then plunking in a big scoop of ice cream. The frothy, sweet goo with a hint of caramel flavor was then ready to be sucked through a straw.

As the name suggests, malted milk is a mix of malt and milk powder. And what is malt? Take some grain, soak it in water until it germinates, dry with hot air, and you have malt. During the process, enzymes are released that break the grain's starch down into sugars such as glucose, fructose, sucrose, and maltose, which are responsible for the malt's sweetness. But the characteristic flavor is due to maltol, a compound that forms when proteins undergo enzymatic degradation. Unfortunately, any malt made from wheat, rye, or barley may contain some residue of gluten and is a no-no for celiacs.

Since malt adds more than just sweetness, the taste of gluten-free Rice Krispies is not identical to the original version. Sugar, still the second ingredient, provides the sweetness, but the taste is now due to whole-grain brown rice instead of refined white rice.

Devoting so much attention to the nuances of a sugary snack whose popularity can be traced not to its taste or nutritional content but to the odd noises it produces may seem frivolous. But if it helps to bring attention to the problems faced by celiac patients and the need for a greater variety of gluten-free products, then the discussion is justified.

Celiac disease is an insidious affliction fully deserving of the description "protean." Proteus was the sea-god of Greek mythology, capable of assuming many forms, just like celiac disease with its variation in signs and symptoms. Gastric problems, depression, irritability, joint pain, mouth sores, muscle cramps, skin rash, tingling sensations, fatigue, and osteoporosis can all be associated with a reaction to gluten. Indeed, Dr. John Weiner, an Australian allergist, suggests that any chronic medical problem that defies diagnosis should be regarded as celiac disease until proven otherwise.

The incidence of celiac disease is roughly one in a hundred, and rising. Several blood tests are available to alert to the possibility of the disease, but a firm diagnosis requires evidence of damage to the lining of the intestine and is only available through a biopsy. The sole treatment is scrupulous avoidance of any contact with gluten. There have even been reports of symptoms being triggered by the use of cosmetics such as body lotions that may have had some ingredients derived from wheat, barley, or rye. Cosmetic manufacturers would do well to take a page from the Rice Krispies notebook and produce a greater variety of gluten-free products.

Celiacs know that they must avoid gluten. But there are many people who may have the disease to some degree without knowing it. In one study, forty subjects who had no symptoms but had a positive blood test were randomly divided into two groups. Half followed a gluten-free diet, the other half ate as usual. After the experiment, those on the gluten-free diet reported an overall feeling of improved health and well-being. A whopping 85 percent elected to continue a gluten-free diet!

Obviously, a case can be made for dietary restriction if a blood test has shown gluten sensitivity. But is it possible that gluten has yet another sinister side? There is more and more talk of "non-celiac gluten sensitivity." People with no evidence of celiac disease claim that avoiding gluten resolves all sorts of symptoms ranging from headaches and bloating to hyperactivity and fatigue.

It seems gluten is fast becoming the latest dietary villain, blamed by some "alternative" practitioners for almost every human ailment. But speculative crackle has to be replaced by evidence before we come to any snap decision and pop off about gluten-free Rice Krispies being the right choice for everyone. Or, considering the high sugar content, for anyone.

Geyser Gets a Little Help From Chemistry

Yellowstone National Park's iconic Old Faithful Geyser is pretty faithful. It can be counted on to erupt every fifty to ninety minutes. Iceland's Great Geysir, from which all other geysers get their name is less reliable. It was mostly dormant for sixty-five years before it began semi-regular eruptions again in 2000, thanks to an earthquake. But in New Zealand, you can set your watch by the eruption of the Lady Knox Geyser, named after a former governor of the country. At exactly 10:15 a.m. every day, a spectacular plume of water and steam bursts into the air to a height of up to 20 meters. How can a natural phenomenon be so predictable? Well, in this case, nature gets a little help from chemistry.

At the appointed time, a detergent solution is poured down the channel from which the water erupts. This has the effect of reducing the surface tension of the water that deep within the shaft has been heating up to boiling temperatures due to underground volcanic activity. Surface tension refers to the attractive force between water molecules, and is in fact responsible for water being a liquid at ordinary temperatures. Liquids are characterized by the close proximity of their component molecules, while in gases the distance between molecules is much greater. If the surface tension of a liquid is decreased, the H_2O molecules can separate from each other with greater ease, with the result that the liquid turns into a gas. Molecules of surfactants, a class of substances that encompasses soaps and detergents, wiggle their way in between water molecules, allowing the boiling liquid to turn instantly into steam. The steam then forces the water that has collected in the channel to burst upwards, and we have an eruption.

The possibility of making a geyser erupt artificially was discovered by accident in New Zealand in 1901, when an "open prison" was established at Wai-O-Tapu for criminals who were deemed not to be a danger to society. It just so happens that this is one of the most volcanically active areas of the world, and the prisoners took advantage of the hot water seeping up from the natural thermal springs to wash their clothes. One day, one of them must have used just the right amount of soap and triggered an eruption when the soap solution found its way down the fissures in the rock into the chambers in which underground water had pooled. This is the concept used today to stimulate eruption of the Lady Knox Geyser, although detergents have replaced soap because they have been found to be less damaging to the geyser's internal natural plumbing. On occasion, Iceland's Great Geysir had also been "soaped," but this is now prohibited for environmental concerns.

Long before the prisoners made their accidental discovery, the science of geyser eruptions had been worked out by none other than Robert Bunsen, of burner fame. Actually, Bunsen did not invent the famous burner, but he did improve upon existing equipment by showing that mixing the combustible gas with just the right amount of air led to a high-temperature non-luminous flame. Such a flame was very useful in the development of Bunsen's most famous discovery, the spectroscope. In collaboration with physicist Gustav Kirchhoff, Bunsen designed an instrument that would pass the light emitted from a sample heated by his burner through a prism. The "spectrum" of light produced was found to be characteristic of the element found in the sample. Before long, Kirchhoff and Bunsen had identified cesium and rubidium as new elements and paved the way to the identification of thallium, indium, gallium, and scandium by others through spectroscopy.

In 1845, during his tenure as professor of chemistry at Marburg University, Bunsen was invited by the Danish government on a geological trip to Iceland following the eruption of Mount Hekla. Bunsen had a lifelong interest in geology and used the occasion to study the gases released from volcanoes and performed analyses on volcanic rocks. He also became interested in Iceland's abundant geysers, especially The Great Geysir that propelled water to a height of some 50 meters. Bunsen hypothesized that eruption occurred when a column of underground water was heated around its middle by volcanic activity. In the true spirit of science, in the laboratory Bunsen constructed an artificial geyser consisting of a basin of water from which a long tube filled with water extended upwards. He heated the tube at various points and showed that it was when the water in the middle reached its boiling point that an eruption occurred, just like in a natural geyser.

Geysers can do more than excite tourists. In Iceland, hot water from geysers is used to heat homes and warm greenhouses, allowing food to be grown in an otherwise inhospitable climate. The accumulation of steam deep within the ground that makes geysers possible can also be tapped by geothermal power plants to produce electricity. Geothermal energy is a great source of electricity, but drawing off the steam can lead to the destruction of geyser activity.

Not all geysers gush steam and hot water. In the case of cold-water geysers, the eruptions are driven by the carbon dioxide gas that forms as limestone, which is composed of calcium carbonate, decomposes. The gas becomes trapped in underground aquifers until it builds up enough pressure to explode toward the surface through cracks in the strata, propelling water into the air. Some of the gas remains in the water in the form of small bubbles, so the geyser actually dispenses soda water.

If you want to experience a mini cold-water geyser, just drop a couple of Mentos into a bottle of Diet Coke. But do it outside. It makes a mess. If you use a special tube (available from Steve Spangler Science) that can simultaneously drop seven to ten Mentos into the bottle, you'll be treated to a true spectacle as the liquid bursts to the stunning height of about 10 meters. That's still some 490 meters short of the super eruptions once produced by the Waimangu Geyser in New Zealand between 1900 and 1904, before the geyser's natural plumbing was destroyed by a landslide. The world's tallest geyser now is Yellowstone's Steamboat Geyser, which propels water some 90 meters into the air. Unfortunately, its eruptions are not predictable. Except on YouTube.

Possums and Kiwis

New Zealand is an amazing place, as I discovered on a lecture tour. Admittedly, I was no great expert on New Zealand before my visit. Sheep, rugby, and kiwi were the words I associated with the country, not possum, weasel, ferret, or rabbit. And I certainly did not connect sodium fluoroacetate with the country. But as I was to learn, New Zealand uses more than 80 percent of the world's production of this chemical.

What we are talking about is a biodegradable pesticide used to control the population of the common bushtail possum. Kiwis, as New Zealanders are known in homage to their national bird, bristle at the very mention of this marsupial. Marsupials are mammals that carry their young in a pouch. Possums tend to feed on the newest shoots produced by vegetation, and as a result, many plants and trees are eventually unable to produce new growth and die. This can have a dramatic effect on biodiversity, since other birds and animals also rely on the same plants for food. Possums are also transmitters of bovine tuberculosis and constitute a threat to beef and dairy industries. And these predators also disturb nesting birds and eat their eggs and chicks. They are partly responsible for the decline in population of the beloved kiwi, an intriguing flightless bird that only comes out of hiding at night. Unfortunately, possums are also nocturnal creatures and a kiwi-possum encounter is not to the bird's advantage.

It turns out that possums are not native to New Zealand, a country that with the exception of two species of bats had no mammals until the middle of the nineteenth century. It was then that European settlers introduced possums and rabbits, with results that would turn out to be catastrophic, but which at the time were unforeseen. Since the country's flora and fauna had evolved with no worries about mammals feasting on them, they had developed no defense mechanisms against the newly introduced animals.

Possums were originally imported into New Zealand to establish a fur trade. But no one could have predicted that, helped by an abundant food supply and lack of predators, their numbers would skyrocket to about 70 million before being reduced by conservation measures. Still, the animals remain a major pest, devouring millions of tons of vegetation every year. The fur industry has been hit by the activities of animal rights groups but it has countered by blending possum fur with merino wool for use in clothing. "Merinomink" or "eco-possum," as the material is called, now accounts for 95 percent of all the fur from possums caught in the wild. This, however, has little impact on the total population.

The use of sodium fluoroacetate-laced bait, however, can reduce possum numbers. Commonly referred to as 1080, its original chemical catalog number, fluoroacetate disrupts the energy producing mechanism in mammalian and insect cells. It is added to carrot or cereal bait usually dropped from helicopters, capable of achieving an impressive kill rate of 98 percent in a targeted area. While commercial fluoroacetate is synthesized by chemical companies, the compound does actually occur naturally in a variety of plants that grow in high-fluoride soils, with black tea leaves from India or Sri Lanka being a prime example. A cup of tea brewed from these leaves will result in a concentration of about 5 parts per billion of fluoroacetate, which is 1.5 times the drinking-water limit. No human health effects have ever been noted with consumption of such tea, and none would be expected with consumption of drinking water that contains trace amounts.

In the environment, 1080 is degraded by soil microbes and fungi into nontoxic substances. That doesn't mean its application is problem free. Dogs can be poisoned, either by eating the bait or the carcass of an animal that has been killed by the chemical. Deer may also be poisoned, but these animals have also evolved into a pest that can cause serious damage to forests.

The introduction of possums to New Zealand is an excellent example of how easily an ecosystem can be upset. Rabbits are another example. They were introduced around the same time as possums to provide game for sportsmen, to serve as a source of food, and to give British settlers a "homey" feeling. Rabbits adapted quickly, and by the 1890s, they were reaching plague proportions. A trade in rabbit fur and in canned and frozen meat developed, but the effects on the land were catastrophic. The furry rodents competed with sheep for grass, with a dozen rabbits eating as much as a single sheep would. They stripped hillsides of vegetation, leaving no covering and leading to erosion when heavy rains lashed the countryside.

With no natural predators, rabbits spread and sheep starved. To combat the problem, ferrets and weasels were imported from Europe. They preyed on rabbits, but they also feasted on the native birds, which had no defense mechanisms and were now totally at the mercy of the new predators. Rabbits multiplied too quickly to be controlled by the weasels and ferrets, and attempts were eventually made to try to reduce their population by infecting them with a virus. Rabbit Hemorrhagic Disease causes blood clots to develop, killing a rabbit within thirty to forty hours from heart and respiratory failure. Their numbers have been somewhat reduced, but rabbits appear be developing an immunity to the disease.

What is needed is a researcher who can pull some anti-rabbit magic out of a hat. And that may well happen because science in New Zealand appears to be very strong, especially in the area of conservation. I was thrilled to see the efforts being made to ensure that future generations will also be able to see penguins waddling from the ocean, albatross majestically soaring through the air, and kiwis pecking at the ground with their long beaks in a constant search for food.

While the penguins and albatross can be admired in the wild, seeing a kiwi in its natural habitat is extremely difficult. The birds only come out during the night and hide very effectively during the day. There are some night excursions with special flashlights offered, but there is no guarantee of an encounter. However, at

Rainbow Springs, a fascinating wildlife preserve near Rotorua, kiwis are raised with great care before being released into the wild. The whole operation can be viewed, including some kiwis that are kept in a very large, nature-like glass enclosure. It is dark with just a few reddish lights and you do have to be patient to see a kiwi scamper by. But it's worth the wait; kiwis are amazing birds. Incidentally, they have nothing to do with the fruit, which was named after them in order to give the "Chinese gooseberry" a New Zealand identity for export purposes.

The golden fruit variety is now threatened by a bacterial infection that may have been introduced by kiwi pollen imported from China to help fertilize local trees, yet another example of the unpredictability of environmental interventions. But for now, golden kiwi is available in stores and it is delicious. And nutritious. Unfortunately, in some people, kiwi can trigger an allergic reaction that can range from itching around the mouth to facial swelling, abdominal pain, vomiting, and breathing difficulties. In rare cases, kiwi reaction can result in anaphylaxis. The reaction is likely due to a protein-hydrolyzing enzyme called actinidin, which can be broken down with heat. Cooked kiwi, therefore, does not cause a reaction. I don't know whether the piece of kiwi that topped the pavlova, New Zealand's national dessert that I tasted, was cooked or not, but I can attest to the fact that the concoction was delicious.

Beep . . . Beep . . . It's Moscow Calling!

The "beep . . . beep . . . beep . . ." sounded innocent enough, but it shook America to its very core. Why? Because it was coming from outer space! No, the military personnel monitoring radio signals did not pick up a transmission from aliens.

This beep was coming from a transmitter placed inside a twenty-three-inch-diameter ball made of aluminum, titanium, and magnesium. A ball that was orbiting Earth, passing over Washington, DC, every hour, emitting an irritating signal that sent a clear message: we are here and you are not! As far as the space race went, it seemed the Soviets were off to a flying start! The date was October 4, 1957.

The race for space had become hot and heavy in the 1950s, with the Americans and Soviets engaged in a furious contest with serious political overtones. The stated goal was exploration of space, to go where no man had gone before. The truth was more down to Earth.

By the middle of the twentieth century, both the Soviets and the Americans had an arsenal of nuclear weapons but lacked a truly effective delivery system. The bombs dropped on Hiroshima and Nagasaki had been carried by airplanes, but those missions had surprise on their side. With the widespread introduction of radar, bombers were unlikely to evade defenses. Missiles were seen to be the ideal delivery system.

But there was a problem. Missiles at the time did not have intercontinental range. That's why the Soviets had their eyes on installing missiles in Cuba and the Americans in Western Europe and Turkey. But clearly a missile that could deliver a nuclear weapon across the ocean was preferable. That, however, required a booster with much more oomph than any that had been developed up to that time.

The Germans had some success with the remarkable V-2, a rocket that terrorized Britain with its conventional payload during World War II. But Britain was only a couple of hundred miles away. A far more powerful rocket would be needed to carry a payload across the ocean. And one would make its appearance on that historic October day in 1957, when the Soviets stunned the U.S. by launching *Sputnik*, the world's first artificial satellite. While this innocuous-looking ball had no great practical importance, it had a huge political and propagandistic impact.

To achieve Earth orbit, a satellite has to be traveling fast enough to cancel out the pull of gravity. This is roughly 28,000 kilometers per hour, or 17,500 miles per hour. That's faster than a speeding bullet! Furthermore, the satellite has to be boosted to an altitude where there is no longer any significant atmosphere to impede its forward motion due to friction with air. That's 160 kilometers, or 100 miles, up. By putting *Sputnik* into orbit, the Soviets had demonstrated that they had a booster rocket powerful enough to achieve the required speed and altitude, which also meant that they had a rocket powerful enough to reach America!

The Soviets followed the launch of *Sputnik 1* just a month later with *Sputnik 2*. This one had more than a radio transmitter inside. It contained a live dog! Laika was the world's first cosmonaut! Putting a man into space had been a long-standing dream, but some important questions had to be answered before risking a human life. How intense was solar radiation in space? How would high-energy cosmic rays affect a living organism? What about the high acceleration of a launch? What would be the effect of weightlessness? Laika, Russian for "barker," would answer some of these questions. Unfortunately, she wouldn't be barking for long. There were no plans to bring Laika back to Earth, and she became the first casualty of the space race.

The U.S. finally managed to place a satellite into Earth's orbit three months later with the launch of *Explorer 1*, and thereby showed that it too now had intercontinental ballistic missile capabilities. Then, in 1960, the Americans achieved their first "first" by orbiting *Echo 1*, a giant, inflated metal-coated ball. This was the world's first communication satellite. Radio and television signals could be reflected off its metal cover to circumvent communication problems caused by the curvature of the Earth. But it didn't take long for the Soviets to cast a shadow over this success.

In 1961, they pulled off another stunning breakthrough by successfully orbiting *Vostok 1*. This outdid the barking dog. The satellite had a talking man inside! Yuri Gagarin completed one full orbit of Earth before *Vostok* was slowed down by a rocket burst that was fired in the direction opposite to its travel, allowing gravity to take over and pull the satellite back to terra firma. Just two hours after being launched into space, Gagarin landed softly by parachute and delivered a hard blow to America's ego.

The U.S. had a rather weak rebuttal to Gagarin's flight a couple of weeks later when it managed to launch Alan Shepard into space on a suborbital flight that lasted only about fifteen minutes. When asked what he thought about while he was sitting in his *Freedom 7* capsule waiting "for this candle to be lit," Shepard remarked he had been reflecting on the fact that "every part of this ship had been built by a low-bidder." Unlike Gagarin, whose flight was controlled totally from the ground, Shepard at least got to exercise some piloting skills. Nine months after Shepard's mission, John Glenn became the first American to orbit the Earth, going around the world three times.

In 1965, the Soviets scored another first when Alexei Leonov exited his orbiting capsule and "walked" in space, a feat soon duplicated by U.S. astronaut Ed White. Both the superpowers were now primed to reach for the moon.

American technology triumphed in 1969, with the "successful landing of a man on the moon and returning him safely to Earth," a project initiated by President Kennedy just after Gagarin's successful mission, and one of mankind's greatest achievements. But that "one small step for man, one giant leap for mankind," could not have been taken without the contributions of numerous scientists and engineers as well as astronauts and cosmonauts of both the human and animal variety. Next step, Mars! But don't hold your breath. Going to the moon is virtually child's play when compared with a mission to Mars. That is not to say it will not happen. American rocket pioneer Robert Goddard said it very well in 1904: "It is difficult to say what is impossible, for the dream of yesterday is the hope of today and the reality of tomorrow."

Be Glad They're Asking About Liquids and Gels

"Do you have any liquids, gels, or powdered fruit drinks?" Except for the powdered fruit drinks, such questions have become routine at airports. But back on July 10, 2006, I had no idea why I was being asked this bizarre question. Why would the agent be concerned about my toiletry and dietary habits? I couldn't make heads or tails of it. The only connection with flight that sprang to mind was with Tang, the orange-flavored crystals that John Glenn took along on his orbital mission in 1962. But we were traveling from London to Budapest and that trip was presumably not going to take us through outer space.

Having just spent a harrowing day at Heathrow queuing for bathrooms and fighting for food after the cancellation of all flights, I gave a curt "no" for liquids or gels, and muttered something about powdered fruit drinks not being my cup of tea. At the time, all I knew was that the massive disruption was caused by some sort of terrorist threat. Only when we got to Budapest did I hear that the threat had something to do with a "liquid bomb." And amazingly, with Tang! My chemical curiosity was of course aroused, but the matter took on a personal touch when I learned that Air Canada's London-Montreal flight, the one we were going to take back home a week later, was one of the ones targeted. Which specific day the terrorists had chosen to try to blast seven planes out of the sky was not clear, but the attacks were apparently imminent, judging by the fact that the terrorists, who had been under extensive surveillance for a month, were suddenly arrested on the eve of July 6.

The whole caper began when British security secretly opened the baggage of Ahmed Ali Khan as he returned from Pakistan. Khan had raised some red flags because of his hard-line anti-Britain political stance, and when his suitcase was found to contain a large number of batteries and a supply of Tang, officials decided to mount a surveillance operation. It seemed unlikely that Khan was into battery-powered gizmos, or that he was bent on cleaning his automatic dishwasher (yes, Tang, because of its citric acid content is great for that), or that he had developed such a fondness for orange-flavored colored water that he had to take a supply of Tang on foreign trips.

After one of Khan's associates was seen disposing of empty bottles of hydrogen peroxide, a video camera was secretly planted and caught the men constructing some sort of device out of beverage bottles. When Khan was seen checking out flight schedules at an internet café, the decision was made to arrest him and his bunch. Of course, details of this operation were not released but somehow reporters got wind of hydrogen peroxide and Tang being involved.

And then the speculation started. Newspaper accounts proposed that the terrorists were actually going to make a bomb from chemicals smuggled through security disguised as beverages by coloring with Tang. Acetone, hydrogen peroxide, and sulfuric acid would be used to make triacetone triperoxide (TATP), a powerful explosive. The necessary materials would not be hard to acquire. Acetone is readily available as nail-polish remover, sulfuric acid is the acid in car batteries, and the concentrated hydrogen peroxide needed can be made by boiling off the water from the 3 percent peroxide sold in pharmacies. Indeed, TATP can be made quite easily by a competent chemist, and it has been used in many a suicide bombing. But synthesis requires careful temperature control, mixing, filtering, and drying, hardly the operations that could be carried out in an airport or airplane toilet.

As more information came to light during the trial of the terrorists, other possible scenarios emerged. Apparently one of the videos taken at the "bomb factory" (as the house where the gang met was dubbed) had shown Khan drilling a hole in the bottom of a bottle with syringes and battery casings nearby. The exact details of what the men were doing and the various chemicals found after the arrest were described to the jury but were not made public.

Speculation was that the fruit crystals dissolved in concentrated hydrogen peroxide were to be introduced by means of a syringe through the bottom of a bottle that had been emptied by the same means. Hydrogen peroxide is an excellent source of oxygen and the sugar in the powdered beverage can serve as a fuel, setting the stage for an explosion. All that is needed is a detonator, which can be made by filling a hollowed-out battery with hexamethylene triperoxide diamine (HMTD). Sounds like a complicated task, but HMTD can be made from hydrogen peroxide, ammonia, and formaldehyde. Khan and his fellow terrorists could have done this.

Supposedly, the idea was to fill a bottle with the explosive mixture, seal the hole at the bottom with Krazy Glue, and take it aboard the plane as a beverage. At the appropriate time, the cap would be removed and the detonator-filled battery shell dropped into the bottle. The explosion would then be triggered with a jolt of electricity from a camera. The jury was in fact shown a video of an explosives expert carrying an orange-colored drink into the mock-up of an airplane fuselage and causing a devastating explosion. What exactly was in the bottle we do not know, because the technical details were only for the eyes and ears of the jury.

The accused admitted to making explosives, but claimed that they were just going to set them off in a public place to make a political statement without causing harm. Defense lawyers pointed out that in fact no flight reservations had been made. But they couldn't explain away the suicide videos found after the arrests and threats of jihad uttered. The jury was convinced of the men's guilt and the major players were sentenced to life in prison.

So, could they have pulled it off? I've looked into the chemistry in much greater detail than I described here, for obvious reasons. Let's just say that the next time you are asked if you have liquids or gels, be glad they're asking.

Smashing Atoms to Smithereens

I set my alarm clock for 1 a.m. so that I could wake up in total darkness. Because only with eyes accustomed to the dark would I be able to "see genuine atoms split"! For weeks I had been waiting for the arrival of "the world's only nuclear powered education toy," and now I was finally ready to look into my "spinthariscope."

Don't picture some complex apparatus. The spinthariscope is more like a cosmetic cream jar with a viewing lens on the top. This curious little device was the invention of Sir William Crookes, best known for the Crookes tube, the forerunner of the vacuum tube. But the British scientist also had an interest in radioactivity, an interest that led to the invention of the spinthariscope, a name he derived from the Greek word "spintharis" for spark. A most appropriate name, since the sparks that could be seen by looking into the spinthariscope sparked a great deal of curiosity.

Crookes's invention made its debut in 1903, at a "conversazione," or gathering, organized by the Royal Society at Burlington House in London. The Royal Society was founded in 1660 with the aim of improving "natural knowledge," as science was then known. To this end, the Society organized these gatherings where the social and scientific elite had a chance to get involved in scientific discussions and view exhibits, some of which were "hands-on." On May 15, 1903, the agenda included poisonous sea snakes from India, hopefully not a hands-on exhibit, and an array of displays by Sir William Crookes on the properties of the emanations of radium. It was here that people first learned about the spinthariscope and the sparks it produced. Soon it would become a popular item among ladies and gentlemen who wanted to show that they were up to date on the latest atomic technology and many a child had his interest in science sparked by the gift of a spinthariscope.

Soon after Marie and Pierre Curie's discovery of radium, Crookes, like many other scientists at the time, became fascinated with radioactivity. It was very much of a mysterious phenomenon at the time, but it was known that exposure to radiation caused zinc sulphide to emit a bluish light. One day, Crookes was experimenting with what was then perhaps the most expensive material on Earth, radium bromide. Inadvertently, he spilled a few grains onto a zinc sulphide screen. As he later described it, "the surface was immediately dotted about with brilliant specks of green light, some being a millimeter or more across, although the inducing particles were too small to be detected on the white screen when examined by daylight." It was a scintillating discovery! In fact, it would lead to the first scintillation counters, instruments that measure radiation by means of tiny visible flashes of light produced when radiation strikes a phosphor, a substance that absorbs energy and then reemits it as light. But before scintillation counters, there was the spinthariscope.

The first ones were beautiful little tubes made of brass, and they produced dramatic results. As Crookes himself described, "on bringing the radium nearer the screen, the scintillations become more numerous and brighter, until when close together, the flashes follow each other so quickly that the surface looks like a turbulent, luminous sea. When the scintillating points are few there is no residual phosphorescence to be seen, and the sparks succeeding each other appear like stars on a black sky."

Crookes thought the effect was due to the bombardment of the screen by the electrons cast off by radium, with each scintillation rendering visible the impact of an electron on the screen. Close, but no cigar! While radium does give off beta radiation, which is really a beam of electrons, it also releases alpha particles (helium nuclei) as it spontaneously decays into radon. It is these energetic alpha particles that cause the flashes of light as they collide with the screen. The spinthariscope actually allowed the direct observation of individual nuclear disintegrations!

Of course, the people who bought the attractive brass tubes didn't really know what they were looking at, but it did give them a feeling of being up to date with the wonderful progress of science. Nor were they aware of the fact that radium was a dangerous material, although it would have posed no risk as long as it was locked inside the spinthariscope. In the 1950s, the scopes enjoyed a modest revival as educational toys, including in some curious formats, such as the Lone Ranger Atom Bomb Ring, a "seething scientific creation." It was available for 15 cents and a boxtop from Kix cereal. Just squint into the secret lens on the Kix Atomic Bomb Ring, and "Zowie, lookit those atoms kick the bucket! See real atoms split to smithereens inside the ring!"

Perhaps now you can understand why I so eagerly anticipated the arrival of my spinthariscope. I didn't really expect to see atoms being split to smithereens. That happens in a nuclear reactor, not in a spinthariscope. But I was expecting to see the results of natural radioactivity. My spinthariscope was powered not by radium, but by a tiny amount of natural thorium ore. Like radium atoms, those of thorium give off alpha particles on their journey toward becoming atoms of radon. There is no risk here; alpha particles cannot even penetrate a sheet of paper and can only travel about an inch in air.

So there I was, bleary eyed, in the middle of the night, staring into the little white jar and hoping to see a dazzling nuclear display. Instead, I was treated to some very, very tiny flashes of bluish light. Technically, one would call them scintillating, but they were hardly worth getting up in the middle of the night for. Still, there is the historical impact. For it was a chance encounter with a spinthariscope that led to Rutherford's classic experiments that resulted in his proposing the nuclear model of the atom. My spinthariscope experience, though, proved to be a little disappointing. Anybody want to trade for a Lone Ranger Atom Bomb Ring?

Justice Full of Beans

Historically, in Western societies, duels were fought with swords or pistols. But the Efik people along the Calabar River in Africa had a different idea. They dueled with beans! Not any old bean. These beans were found in the pods of a plant the Efik called "esere." One would be cut exactly in half, and the pieces swallowed by the adversaries. Supposedly, the man who had justice on his side would live, and his opponent would die. In most cases, the poison in the bean, eventually named "eserine" or "physostigmine," took care of both men.

The Efik also used extracts of the Calabar bean in "trials by ordeal," as described by British missionary William Daniell in an 1846 report to the Ethnological Society of Edinburgh: "The condemned person, after swallowing a certain portion of the liquid, is ordered to walk about until its effects become palpable. If, however, after the lapse of a definite period, the accused should be so fortunate as to throw the poison from off his stomach, he is considered innocent, and allowed to depart unmolested." If the accused didn't survive, he was deemed to have been guilty and justly punished.

There actually may have been some rationale to the madness. A guilty person likely hesitated in swallowing the liquid, keeping it in his mouth as long as possible. An innocent victim would have swallowed it quickly, eager to prove that he had been wrongly accused. It turns out that physostigmine is readily absorbed through the mucous membranes of the mouth, causing rapid death, but on entering the stomach, it triggers a vomiting reflex and is quickly expelled! The same procedure was, and possibly still is, used in some parts of Africa to reveal witchcraft. A real witch will foam at the mouth and collapse, while one falsely accused will regurgitate the noxious brew and survive.

This sort of chemical witchcraft invited investigation by Europeans, whose appetite for plant extracts had been whetted by the isolation of quinine from the bark of the Peruvian cinchona tree, curare from a South American vine, strychnine from the seeds of the East Indian nux vomica tree, atropine from belladonna berries, and digitalis from the foxglove plant. Each of these had the capacity to serve as a poison or a drug, depending on the dose. Scottish physician Sir Robert Christison, who would later become president of the British Medical Association, was especially intrigued by the Calabar bean. A true pioneer of research, he performed an experiment that today would be unthinkable. He used himself as a guinea pig!

Starting with an eighth of a seed, Christison ingested larger and larger pieces of the bean. With the smallest dose, he noted only a slight numbness of the extremities, but as he increased his intake, he became giddy and took to bed with a feeble pulse. Time and a strong cup of coffee eventually restored him. There's no evidence he pursued this foolhardy experiment any further, but Christison noted he had felt no pain and suggested "that the drug may be humanely employed in the execution of criminals condemned to death."

Thomas Fraser, one of Christison's assistants, explored Calabar extract further, although not in his mentor's cavalier fashion. Fraser noted that when used as an eyedrop, the extract caused tearing and a contraction of the pupil. This was exactly the opposite of what was seen with belladonna extract! Could it be that the Calabar bean could serve as an antidote to atropine, the belladonna poison? That did indeed turn out to be the case. Many a child poisoned by the accidental ingestion of deadly nightshade berries has been saved by the timely administration of physostigmine. The same goes for thrill-seekers who were brought down as they sought a high in the atropine-containing seeds of the plant known as Angel's Trumpet. Physostigmine was also found to effectively reverse the effects of curare, a substance that, like atropine, had a paralyzing effect on muscles.

The mechanism of action of physostigmine was finally unraveled in the 1920s, thanks mostly to the efforts of Austrian pharmacologist Otto Loewi. It turns out that the drug blocks the action acetylcholinesterase, an enzyme that degrades acetylcholine, a key neurotransmitter needed to relay information from nerve to muscle cells. If the enzyme's activity is impaired, more acetylcholine is available to stimulate muscle contractions.

Of course, victims of atropine or curare poisoning are not treated with Calabar beans; they are given an appropriate dose of pure physostigmine, first isolated back in 1864. By 1935, Percy Julian in the U.S. had managed to synthesize the compound, making a ready supply available to researchers. Before long, it was also found to reduce excess pressure in the eye, making physostigmine the first effective drug against glaucoma.

And then, in 1934, came the "miracle of St. Alfeges." Dr. Mary Walker at St. Alfeges Hospital in London had noted that the symptoms of myasthenia gravis, a devastating neurological disease, were similar to the effects of curare. Since physostigmine reversed curare toxicity, she wondered if it might be useful in the treatment of myasthenia gravis. Injection of a patient with the drug led to a rapid, albeit temporary, response. Eventually, synthetic derivatives of physostigmine would become the cornerstone in the treatment of myasthenia gravis. Since Alzheimer's disease is also characterized by an acetylcholine deficiency, physostigmine and its derivatives are also being explored in its treatment. And since acetylcholine also plays a role in stimulating ejaculation, men suffering from spinal cord injuries have been able to provide sperm and father children after being injected with physostigmine.

Drug discovery can certainly take some circuitous routes. Who could have ever guessed that attempts in Africa to unmask witches would lead to treatments for neurological diseases, or to the recognition that physostigmine and atropine were mutual antidotes. Imagine if one of our African duelists had known this.

A chaser of a belladonna berry after downing the Calabar bean would have assured victory. Only if the dose was right, that is.

Of Mice and Men and Apples and Oranges

The cartoon would be meaningless without the accompanying article. There's Eve standing under a tree, being offered an apple by the serpent. "No thanks," she says, "I'd rather have a tangerine." A glance at the headline above the article immediately makes sense of Eve's quip. It's pretty catchy: "Tangerine a Day Keeps Heart Surgeons Away." Is this evidence-based science or is it the brainchild of an overly enthusiastic headline writer?

Both the headline and cartoon were sparked by some elegant work at the University of Western Ontario focusing on nobiletin, a naturally occurring chemical found in tangerines. It belongs to a family of compounds called methoxylated polyflavones, which have garnered a great deal of chemical interest because of their potential antioxidant, anti-inflammatory, anti-carcinogenic, and cholesterol-lowering activity.

Dr. Murray Huff and his team fed two groups of mice a calorie-rich, high-sugar diet designed to make them fat and prone to heart disease. But one of the groups also had a daily dose of nobiletin mixed into their food. And what a difference that made! The mice on the nobiletin-laced diet put on less weight, and even more importantly, had lower blood levels of insulin, glucose, triglycerides, and cholesterol. They were also less likely to develop fatty liver and had less of a buildup of plaque in their arteries! This news led to the headlines about tangerines reducing heart disease risk. This despite the fact that tangerines were not even involved in the study! But I suppose that "huge doses of purified methoxylated polyflavones increase insulin sensitivity and attenuate atherosclerosis in mice" is not as catchy a headline as "Superfruit Tangerines Can Reduce Heart Attack Risk." It is, however, more realistic.

The first point of interest in any such study is the dose of the chemical that was used, which in this case was roughly 10 milligrams of nobiletin per day. A tangerine contains about 1 milligram, with virtually all of it in the peel and the pith. That's the white stuff that separates the peel from the fruit, which most people discard. The portion we eat contains hardly any nobiletin. So there's absolutely no justification for the headline about a tangerine a day keeping the heart surgeon away.

The authors of the paper, of course, don't make any such suggestion at all. They correctly conclude, "the use of nobelitin provides insight into potential targets for the treatment of abnormal lipoprotein and glucose metabolism, characteristic of insulin-resistant states and premature atherosclerosis." If any benefit is to be had from nobiletin, they say, it is likely to come in the form of a supplement. Actually, similar products are already on the market, and since they are derived from a natural source, do not require a prescription. Sytrinol claims to be a mixture of polymethoxylated flavones extracted from citrus and palm fruits along with other "proprietary ingredients." According to the manufacturer, it lowers cholesterol by up to 30 percent. But there is no exact description of the ingredients, and the studies have not appeared in the peer-reviewed literature. Still, it may be one of those hyped "all-natural" products that actually has a chance of working.

So, if "a tangerine a day keeps the doctor away" amounts to no more than a headline writer's fancy, what about the age-old "apple a day" adage? I've always found that turn of phrase appealing, and even used it as a title for one of my books. I was, however, using it as a metaphor for eating more fruits and vegetables, rather than as a literal prescription for health. Because of my apple connection, I was intrigued by the recent appearance of a brand-new slew of headlines about how "an apple a day keeps heart disease away." Here we go again, I thought, some lab finding or animal study of borderline significance to humans has provided fodder for the headline writer's cannon. But I was wrong. Bahram Arjmandi's study at Florida State University may not exactly justify the headlines, but it is relevant to us.

First, the study actually used human subjects in sufficient numbers and over a sufficiently long period of time. And they were not asked to consume some outrageously unrealistic amount of apples. The first group, made up of 160 post-menopausal women, was asked to eat 75 grams of dried apples a day, roughly equivalent to two fresh apples. A control group consumed an equivalent amount of prunes (dried plums). After a year, total blood cholesterol in the apple group decreased by 14 percent, and LDL, the notorious "bad cholesterol," dropped by 24 percent. No equivalent results were seen with prunes. In addition, lipid hydroperoxides, a measure of free radical activity, declined by 33 percent, and levels of pro-inflammatory C-reactive protein went down by a similar amount. And to everyone's surprise, not only did the extra 240 calories a day not cause weight gain, but there was also an average weight loss of 1.5 kilograms.

Just which apple components are responsible for these effects isn't clear. Pectin, a form of soluble fiber commonly used as a jelling agent in fruit preserves, is a strong candidate for the cholesterol-lowering effect. It can bind bile acids in the gut, preventing them from being reabsorbed, thereby forcing more bile acids to be formed from stored cholesterol. The net effect is a reduction in blood cholesterol. Pectin's ability to lower cholesterol has been clearly shown in pigs and in small-scale human trials. There is even anecdotal evidence of people reducing their cholesterol with a couple of teaspoons of Certo fruit pectin a day. As far as the free-radical scavenging and anti-inflammatory effects go, they are probably to be found in the apple's content of polyphenols.

In the end, comparing the overly exuberant headlines about tangerines and apples is like comparing apples to ummm . . . oranges. The tangerine headline is totally unrealistic, while the apple headline is more reflective of the study that prompted it. Alas, there was no cartoon. There could have been. Here's my proposal: Eve responds to the serpent's offer of an apple: "One? Just one? Evidence-based science says two!"

Imported Fruit May Harbor Terrorists

"Do you have any fruits or vegetables?" the U.S. customs agent asked. "Only my apple a day," I semi-jokingly replied. "I never go anywhere without it." "Well," came the serious reply, "you'll be going without it today!" Foreign apples, I was told, cannot be taken into the U.S. Do Americans have some irrational fear of apples somehow being used by terrorists, I wondered? In a sense, yes, they do. But the fear isn't irrational. The "terrorists," however, are not of the human variety. They're innocent-looking little bugs. But they strike terror into the hearts of farmers. And their eggs can stow away aboard fruit.

Mention "Mediterranean fruit fly" to fruit or vegetable growers in Florida or California, and hearts skip a beat. Slightly smaller than the common housefly, the "medfly" is one of the most destructive fruit pests in the world. The adult female can pierce the skin of any of over 250 different types of fruit and deposit her eggs, which then hatch into little worms called larvae. Also known as maggots, they eat the pulp of the fruit, turning it into a soft mush, often without altering the outside appearance. The fruit then drops from the tree prematurely, allowing the larvae to crawl into the soil. There they transform into pupae that mature into adults and emerge from the soil to mate and start the cycle again. Since the whole cycle can be completed in twenty to thirty days, the scope of destruction can be extremely severe. Entire crops can be wiped out in a matter of weeks. That, of course, has implications not only for the farmer, but for all of us. Much of the fruit we consume, and by all accounts we should be consuming multiple servings a day, comes from either Florida or California. Should the medfly become permanently established there, we will all suffer the consequences. Luckily, so far, while there have been infestations, massive eradication programs have managed to prevent the fly from becoming a permanent pest.

The bug is believed to have originated in Africa, from where it migrated aboard fruit to Europe, the Middle East, Australia, South America, and Hawaii. In the 1930s, the flies found their way to Florida, where panicked agriculture researchers used a spray of molasses laced with arsenic to attract and dispatch them. Subsequent infestations occurred in Florida, Texas, and California, necessitating eradication programs that involved aerial spraying with malathion, an organophosphate pesticide. This was not without controversy because of concerns about people being exposed to a potentially hazardous pesticide. But the spraying did manage to wipe out the flies.

Today, farmers in susceptible areas are in constant fear that the bugs will reappear. Widespread monitoring programs have been established to nip any infestation in the bud. They make use of pheromones, chemicals released by both male and female medflies to attract mates. Medfly pheromones have been identified and can be used to lure insects into traps. Finding a single fly can initiate an immediate quarantine of produce from the surrounding area and the stripping of fruit from all trees in the vicinity. This may be followed by spraying insecticide mixed with compounds that, to the medfly, signal the prospect of a gustatory feast. Spinosad, derived from a soil-dwelling bacterium, is the insecticide of choice, being a safer alternative to malathion, which is reserved as a last resort. Spinosad can also be sprayed under host trees if fruit fly larvae are detected in the soil.

Mass trapping of both male and female medflies is another approach, one that is attractive because it eliminates concerns about spraying. Extensive research has explored the types of traps, the best attractants to use, and the best ways to disperse the attractants. Traps are usually colored yellow, that being the medfly's favorite hue. As far as bait goes, pheromones can be effective for monitoring, but on a large scale, it is not the prospect of sex, but rather the lure of a tasty meal that attracts the most flies. Females favor ammonium acetate, trimethylamine, putrescine, cadaverine, and n-methylpyrrolidine, either alone or in various combinations. This is somewhat curious, since these compounds are more characteristic of rotting meat than fruit. There's no accounting for taste. Male medflies are drawn to trimedlure, a synthetic compound that was found to be effective through screening tests.

Trapping can be effective, especially when used in combination with the Sterile Insect Technique. Millions of medflies are raised in captivity and sterilized by exposure to small doses of radiation. Just before being released, they're exposed to pheromones to make them hungry for sex. The excited flies then fervently seek out their wild brethren, but the union produces no offspring. Since female flies mate only once in a lifetime, this constitutes an effective method of birth control.

Legions of sterilized medflies have recently been released in Broward County in Florida, where finding a few medflies in a monitoring trap has caused alarm. We'll see whether all the scientific ingenuity that has gone into medfly research pays off.

The prospect of the fly becoming permanently established is horrific and would amount to billions of dollars in crop losses a year. And it isn't only growers who would be affected. A medfly infestation causes retail prices to jump and results in less fruit consumed, and consequently, a poorer diet.

Fruit flies do, however, have a redeeming feature. Because of their rapid reproduction, they can be used in aging research. Recent studies have shown that some varieties of fruit flies reared on a diet fortified with polyphenols extracted from apples not only walked and climbed with greater ease as they got older, but lived on average 10 percent longer! Another argument for "an apple a day." At least, for the health of fruit flies. For our health, though, we want to keep these little beasties, particularly the Mediterranean species, from frolicking in southern orchards. If it means taking apples from travelers, so be it. Of course, you can always eat your apple before you approach the customs agent.

The Mesmerizing Power of Belief

To "mesmerize" is to enthrall. And that is exactly what Viennese physician Franz Anton Mesmer did to the patients who flocked to his healing salon in the middle of the eighteenth century. Mesmer's "healing" was based on his belief that the universe was permeated with an invisible fluid that connected people to the planets and to each other. The motion of the planets, he suggested, influenced the fluid, which in turn influenced people's health. Accordingly, "influenza" was a disease attributed to the shifting of heavenly bodies.

Long before Mesmer, Paracelsus, the famed sixteenth-century physician and alchemist, had philosophized about a universal fluid to explain what he believed were changes in the body that reflected changes in the solar system. Essentially this was an attempt to rationalize a belief in astrology. Paracelsus claimed that the universal fluid had magnetic properties and could be guided into afflicted parts of the body by magnets. He based this view on his personal observation of "healing" patients by passing a lodestone, which is a naturally occurring magnet, over their bodies. Unbeknownst to him, Paracelsus was witnessing the powerful mind-body connection that eventually came to be called the placebo effect.

Paracelsus's ideas carried significant weight. Even Isaac Newton contemplated the idea that the universe was permeated with an "aether" that allowed for the transmission of light, gravity, and magnetism. So by the time that Mesmer appeared on the scene, the idea of some sort of cosmic magnetism was quite firmly entrenched. And when Mesmer witnessed one of his mentors, Viennese Jesuit Maximilian Hell, carry out apparent healings by applying steel plates to the bodies of the ill, he concluded that magnetic healing was the wave of the future for medicine. This was an attractive idea for patients because sitting around and gripping magnetized metal rods connected to a tub filled with iron filings was preferable to purging and bleeding, the common conventional medical treatments of the time.

Mesmer went on to explore other means of restoring the body's magnetic fluid and found that having patients sit with their feet in magnetized water while holding cables attached to magnetized trees worked well. It was a classic case of coming to the wrong conclusion based on a correct observation. Patients really did report feeling better, which of course is not the same as being better. But the effect didn't have anything to do with magnetized water or trees, neither of which can be magnetized.

But, based on the positive reports from his patients, Mesmer became convinced that illness was indeed caused by some sort of depletion in an invisible magnetic fluid, and that restoration of the fluid was curative. He assumed that healthy people were permeated with the fluid and began to wonder whether they could transmit some of their excess to the afflicted the same way that it was transmitted from magnets, which he assumed were loaded with it. Before long he discovered that indeed the healing fluid could be passed from the healthy to the ill, prompting him to hatch the theory of "animal magnetism." He chose the term "animal" from its Latin root "animus," meaning "breath," to imply that this was some sort of life force that was possessed by all creatures with breath, namely humans and animals. This force could be passed from the healthy to the ill either by direct contact or just by being in close proximity. Even this idea was not novel.

Long before, the ancient Chinese had spoken of some sort of life force called "qi," which traveled through the body's energy channels, and Hindu culture featured "chakras," which were believed to be some sort of intangible energy centers in the body. Today, practitioners of "reiki" and "therapeutic touch" rationalize their efforts by claiming an ability to manipulate the body's energy field, sometimes described as an aura. This sounds very much like Mesmer's transference of animal magnetism. Modern science has found no evidence for any sort of energy centers or channels or invisible fluids, but there is no question that many patients claim to have experienced positive effects after having such seemingly non-existent entities manipulated. It is likely that the thread that ties all these healing modalities together is the power of belief.

Mesmer himself was forced out of Vienna by a jealous medical establishment that was losing patients to the newcomer. He then set up shop in Paris, catering mostly to wealthy hypochondriacs. Dressed in a long robe embroidered with astrological symbols, he made for an imposing sight as he stared into the eyes of patients and triggered reactions ranging from sleeping to dancing and even convulsions. All these are familiar to scientists who study hypnosis, which is essentially what Mesmer was practicing. Good-looking young men were hired as assistants to sit knee-to-knee with ladies in order to massage whatever ailment they had out of their bodies. Sometimes the curative work was continued in private rooms where the subjects could experience personal satisfaction without guilt.

Like in Vienna, French physicians felt threatened by Mesmer's antics and convinced Louis XVI to set up a Royal Commission to investigate Mesmer and his cosmic fluid. The king was keen because he was not pleased that the queen, Marie Antoinette, had fallen under Mesmer's spell. Mesmer refused to cooperate with the commission, but at the behest of Benjamin Franklin who, along with chemist Antoine Lavoisier and physician Joseph Guillotin, was a member of the commission, an experiment was designed that may well have been the first placebo-controlled trial ever conducted. Blindfolded patients were shown to respond to a non-magnetized tree as well as to one that was magnetized by Mesmer's methods. The committee's conclusion was that "the imagination without the magnetism produces convulsions, and the magnetism without the imagination produces nothing."

Mesmer left Paris and died in obscurity in Switzerland in 1815, but he maintains a prominent place in medical history for having stumbled upon and popularized the power of imagination in influencing health. Interestingly, around the same time, Samuel Hahnemann came up with the theory of homeopathy, another practice that has no physiological basis. The effect on patients was the same as mesmerism, although the explanation of the effects was totally different. Both Mesmer and Hahnemann were unwitting pioneers of the power of the placebo.

IN THE END

Our Posthumous Footprint

They say you can't take it with you. Actually, that isn't quite true. Your earthly possessions stay behind, but there is something that you do take with you. Your body! And decisions have to be made about what is to become of it. Burial and cremation are the traditional choices, but now there is another option on the horizon. A "green" option. You can be resomated. In technical terms, your remains can be subjected to "alkaline hydrolysis." In somewhat less elegant language, you can be washed down the drain.

Why should anyone consider being hydrolyzed? It doesn't sound particularly appealing, but, on the other hand, being consumed by maggots and bacteria, or being set ablaze and turned into ash, have no particular charms either. But resomation leaves less of an environmental footprint. There's no concern about embalming chemicals such as formaldehyde leaching into the water table or mercury from dental fillings being spewed into the air by energy-guzzling crematoria. Cremation requires a temperature of about 1,000°C, which means that a lot of fuel has to be burned, and that means a good dose of carbon dioxide is released. Then there is the problem of mercury from dental amalgams that has a number of European countries already requiring the filtering of mercury emissions from crematorium smokestacks. Resomation uses much less energy than cremation and dental amalgam remnants are easily separated from the remains.

Let's take a look at the science involved in resomation. We've seen horror movies where a body is dumped into a vat of acid only to emerge as a skeleton after a few minutes of ferocious sizzling. That is, let us say, poetic license. But flesh can dissolve in acid. It just takes a bit of time. Actually, "dissolve" is not exactly the right term. "Decompose" is more appropriate. Proteins break down into peptides and individual amino acids, while fats are converted to fatty acids and glycerol. The end result is more of a sludge than a solution, but the body as such does disappear.

The notion of "dissolving" a body in acid first came to the public's attention in 1896, when Herman Mudgett, alias Dr. Henry H. Holmes, was convicted of killing at least twenty-seven people for the purpose of providing skeletons to medical schools. He used copious amounts of acid to remove all traces of flesh from the bones. And just a year later, in a highly celebrated case, Adolph Louis Luetgert, known as the "sausage king of Chicago," was convicted of murdering his wife and disposing of her body by dissolving it in acid.

Curiously, 1896 also saw the publication of Melville Davisson Post's short story "The Corpus Delecti," in which a murderer dumps a victim's body into sulfuric acid and escapes conviction because no body is found. Did this story germinate an idea in Mudgett's or Luetgert's head? We'll never know. But it seems the idea of "no body, no conviction" was firmly in the mind of John George Haigh, perpetrator of one of the most bizarre mass murders in history.

Haigh killed at least six people in the 1940s, apparently to secure their possessions, and got rid of their bodies by dumping them into vats of acid. When arrested, confident that he could not be convicted in the absence of remains, he arrogantly admitted to his crimes. He even described how his true motivation was a need to drink the blood of his victims, inspiring the British press to label him the "vampire killer." Whether Haigh was literally bloodthirsty, or concocted the story hoping for an insanity defense isn't clear, but what is clear is that he was overconfident about no remains being found. A forensic team sifting through Haigh's acid sludge found dentures, gallstones, and part of a left foot. He was sentenced to death by hanging.

More recently, in the 1980s, Mafia boss Filippo Marchese had his enemies disposed of in vats of acid in his Palermo home. Justice sort of prevailed when he suffered the same fate in a revenge killing. Mexican drug cartel hit men have also immersed victims in acid in a bid to dispose of evidence. But, as has been shown by forensic researchers using pig carcasses, it is impossible to erase all traces of a crime in this fashion.

There is no doubt that flesh can be decomposed with acid, but this is actually not the best way to go about the grisly process. It is more effective to use a concentrated base instead of acid for breaking down proteins and fats. This is alkaline hydrolysis, and is the principle of resomation.

A resomator is a sophisticated piece of equipment that converts a human body into an oily liquid and a white powder. It looks like an elongated washing machine, but instead of clothes and detergent, it is loaded with a concentrated solution of potassium hydroxide and a body wrapped in silk. Pressurizing the chamber makes it possible to heat the contents well above the boiling point of water. Essentially, what we have here is a pressure cooker that basically does the same job as the common kitchen appliance. In cooking, the point is to break down complex molecules to simpler compounds. That's just what the resomator does.

Alkaline hydrolysis decomposes the body to a brownish liquid of amino acids, peptides, fatty acids, sugars, and salts. Suspended in the oily liquid are the remains of the skeleton, which can be separated and easily crushed into a white dust consisting mostly of calcium phosphate. Both the liquid and the bone remains can be used as fertilizer, or if desired, the dust can be placed in an urn and returned to the family. Cost, though, is an issue, with a resomator going for nearly half a million dollars. So far, this technology is mainly used for medical school cadavers and animals, but if it gets approval as an alternative to burials or cremation, environmentally conscious consumers may be willing to spend a little extra to reduce their posthumous footprint. In the future, it may be the way to go.

INDEX